Buffer Solutions
The Basics

R. J. Beynon

Department of Biochemistry and Applied Molecular Biology
UMIST, Manchester

and

J. S. Easterby

Department of Biochemistry
University of Liverpool

IRL PRESS
—at—
OXFORD UNIVERSITY PRESS
Oxford New York Tokyo

1996

Oxford University Press, Walton Street, Oxford OX2 6DP

Oxford   New York
Athens   Auckland   Bangkok   Bombay
Calcutta   Cape Town   Dar es Salaam   Delhi
Florence   Hong Kong   Istanbul   Karachi
Kuala Lumpur   Madras   Madrid   Melbourne
Mexico City   Nairobi   Paris   Singapore
Taipei   Tokyo   Toronto
and associated companies in
Berlin   Ibadan

Oxford is a trade mark of Oxford University Press

Published in the United States
by Oxford University Press Inc., New York

© Oxford University Press, 1996

*All rights reserved. No part of this publication may be
reproduced, stored in a retrieval system, or transmitted, in any
form or by any means, without the prior permission in writing of Oxford
University Press. Within the UK, exceptions are allowed in respect of any
fair dealing for the purpose of research or private study, or criticism or
review, as permitted under the Copyright, Designs and Patents Act, 1988, or
in the case of reprographic reproduction in accordance with the terms of
licences issued by the Copyright Licensing Agency. Enquiries concerning
reproduction outside those terms and in other countries should be sent to
the Rights Department, Oxford University Press, at the address above.*

*This book is sold subject to the condition that it shall not,
by way of trade or otherwise, be lent, re-sold, hired out, or otherwise
circulated without the publisher's prior consent in any form of binding
or cover other than that in which it is published and without a similar
condition including this condition being imposed
on the subsequent purchaser.*

A catalogue record for this book is available from the British Library

Library of Congress Cataloging in Publication Data
(Data available)

ISBN   0 19 963442 4

Typeset by R. J. Beynon
Printed in Great Britain by
The Bath Press, Bath

Preface

In almost every experiment that we conduct, we use buffers to control the concentration of an ion that is probably the most reactive of all ions in our experiment—namely the hydrogen ion or proton. Usually we indicate its concentration by pH. Often, the decision as to which buffer compound and conditions to use is already made when we follow existing methods. In other circumstances, we will need to make decisions about the best buffer compound, the concentration at which it will be effective, the ionic strength of the buffer and the temperature at which it will be used. All of these factors are interdependent, and there is a need for a basic understanding of their interactions.

This book adopts a pragmatic approach to the use of buffers for pH control in aqueous solution. It presents the information that is necessary to understand how buffers function, and shows how it is used to make choices of experimental systems. The presentation is rather different from many formal treatments of buffers' behaviour, emphasizing the practical outcomes rather than the detailed theory.

The main themes of the book include a recapitulation of basic concepts, a discussion of acids and bases and their strength, and a treatment of the inter-relationships between pH, ionic strength, and temperature. There is a discussion of the use and care of pH electrodes and meters and of the practical aspects of buffer design and preparation. The book describes software that can be obtained, free of charge, for design of new, correctly calculated buffers and includes a detailed listing of the most common buffer compounds used in the molecular life sciences.

If you do experiments, you need to know how buffers work. This book provides, in an approachable style, the basic information that will give you this understanding.

Manchester	R.J.B.
Liverpool	J.S.E.
June 1996	

Contents

CHAPTER 1 Basic concepts
1. What are buffers? — 1
2. Why do we need to worry about buffers? — 2
3. Why do we need to know how buffers work? — 3
4. A word about 'equations' — 3
5. How to use this book — 4
6. The range of proton concentrations — 4
7. The pH scale — 4
8. Why is pH 7.0 'neutral'? — 5
9. The hydronium ion — 6

CHAPTER 2 Acids and bases
1. Introduction — 7
2. Definitions — 7
   2.1 Strong and weak acids and bases — 8
   2.2 The role of water — 10
3. The pH of a solution of a weak acid or base — 11
4. Structure and acidity — 13
   4.1 Ionizable groups in weak acids and bases — 14
   4.2 Inductive and electrostatic effects — 14
   4.3 Mesomeric effects — 15
   4.4 Steric effects — 16
   4.5 Statistical effects — 16
   4.6 Hydrogen bonding — 17
   Further reading — 17

CHAPTER 3 Theory of buffer action
1. Introduction — 18
2. Weak acids and bases resist pH changes — 19
3. How good is a buffer? – β values — 25
4. pK_a is affected by the solution conditions — 25
   4.1 Activity and concentration — 25
   4.2 Ionic strength — 26

4.3 Ionic strength depends on the pH of a buffer 28
  4.4 Ionic strength influences the pK_a 29
  4.5 Dilution effects 30
  4.6 Effect of temperature 31
 5. Thermodynamic and apparent pK_a values 32
 6. Polyprotic buffers 32
 7. 'Mixed' buffers 33
    Further reading 34

CHAPTER 4 Measuring pH

 1. Introduction 35
 2. The electrode 35
    2.1 The glass electrode 35
 3. Care of a pH electrode 36
    3.1 Keep the electrode filled 36
    3.2 Mechanical abuse 36
    3.3 Contamination of the electrode 36
 4. Replacing an electrode 37
 5. The pH meter 37
 6. Calibration of a pH meter and electrode 39
    6.1 Temperature effects 39
    6.2 Calibration of a pH meter 39
    6.3 Slope adjustment 40
    6.4 A common source of error 40
 7. Alternative electrode technologies 41
    7.1 How can we improve on the glass membrane? 41
    7.2 How can difficult or small samples be measured? 42
    7.3 Resistance to contamination by biological fluids? 42
    7.4 Electrical characteristics of the glass electrode 42
 8. The solid-state electrode 43
    Reading list 44

CHAPTER 5 Preparation of buffer solutions

 1. Where do I start? 45
 2. Following existing recipes 45
 3. Designing a new buffer system 46
    3.1 At what pH is the buffer needed? 46
    3.2 Which buffer? 47
       3.2.1 Charge 47
       3.2.2 Metabolic activity 47
       3.2.3 Insoluble salt formation 48

	3.2.4 Chelation of metal ions	48
	3.2.5 Ultraviolet absorbance	48
	3.2.6 Buffers for chromatography	48
3.3	What concentration of buffer?	49
3.4	What ionic strength?	49

4. Examples of buffer design — 50
    4.1 Overall strategy — 50
    4.2 A simple buffer with control of ionic strength by added salt — 52
        4.2.1 Introduction — 52
        4.2.2 Method — 52
        4.2.3 Notes — 52
    4.3 Simple buffer with no external control of ionic strength — 53
        4.3.1 Introduction — 53
        4.3.2 Method — 53
        4.3.3 Notes — 53
    4.4 A buffer prepared at a different temperature — 54
        4.4.1 Introduction — 54
        4.4.2 Method — 54
        4.4.3 Notes — 54
    4.5 A series of buffers at different pH values using a single buffer species — 55
        4.5.1 Introduction — 55
        4.5.2 Method 1 — 55
        4.5.3 Method 2 — 55
        4.5.4 Notes — 55
    4.6 A set of buffers covering a wider pH range, using different buffer species — 56
        4.6.1 Introduction — 56
        4.6.2 Method — 56
        4.6.3 Notes — 56

5. Practicalities of buffer preparation — 57
    5.1 Which buffer compounds? — 57
    5.2 Water of crystallisation — 58
    5.3 Use of concentrated stock solutions — 59
    5.4 Handling buffer substances — 60

6. Correct buffer descriptions — 62

Further reading — 62

CHAPTER 6 Automatic buffer calculations

1. Automating buffer calculations — 63
2. Macintosh Hypercard 'BufferStack' — 63
3. Buffer calculations on the World Wide Web — 65
4. MS-DOS/Windows software — 66
5. Obtaining the software — 66

Further reading — 66

APPENDIX 1 Properties of common buffer

1. How to use this appendix   67
   Further reading   67

APPENDIX 2 Standards for pH calibration

1. How to use this appendix   83
   1.1 Buffer A: potassium hydrogen phthalate   83
   1.2 Buffer B: potassium sodium phosphate   83
   1.3 Buffer C: sodium tetraborate   84

Glossary   85

Index   87

Basic concepts

1. What are buffers?

- Definition of buffers
- Why understand them?
- How to approach the maths
- Why pH 7.0 is 'neutral'

In 1900, Fernbach and Hubert, studying the enzyme amylase, noted that a partially neutralized solution of phosphoric acid acted as a 'protection against abrupt changes in acidity or alkalinity: the phosphates *behave as a sort of buffer*'. This definition is unchanged today, although we now have a better understanding of the phenomenon, and the reasons why it is important to make use of it.

Let us assume for the moment that you are familiar with the pH scale to express acidity or alkalinity of a solution, and that you know that in aqueous systems pH 7.0 is neutral, solutions with pH values below 7 are acidic, and solutions with pH values greater than 7 are alkaline.

It is easy to demonstrate the phenomenon of buffering (Figure 1.1). We could take 1 ml of a 1 M solution of a strong acid, such as hydrochloric acid, and add it to a litre of pure water. The pH would drop (to a close approximation) from 7 to 3, a decrease of 4 pH units. If instead of pure water we had used one litre of a solution of 0.1 M sodium phosphate at the same initial pH of 7, the pH would only drop by 0.02 pH units to 6.98. Thus, the phosphate solution 'buffers' pH changes. A buffered solution resists changes in pH when the solution is exposed to acids or alkalis that would otherwise cause dramatic changes in pH. No buffer can resist these changes completely; at best they can minimize the effects.

Obviously, if we were conducting an experiment that needed to be kept at a constant pH (which, in practice, is true of most experiments), it would be foolish to dissolve all the reactants in pure water. In the absence of a buffer, any small change in the content of acid or alkali would cause a large pH change. But where would this acid or alkali come from? There are several sources—carbon dioxide dissolved from the air will cause the pH to drop, the reactions that are taking place during the experiment can either produce or remove acid, and actively metabolizing cells can release acidic metabolic end products. Sometimes stock reagents have to be added from acid or alkaline solutions, and these reagents will themselves cause a shift in pH.

In the example above, the sodium phosphate acts as a buffer, the consequence of which is that the solution is able to resist the pH shift that would otherwise be caused by a substantial addition of acid. So why can't we just use sodium phosphate, or 'Tris', or 'acetate', or any of the others that we come across, in every single experiment? Mostly because each buffer is only effective over a rather narrow pH range, and we have to use different buffers at different pH values. Even then, there are still several

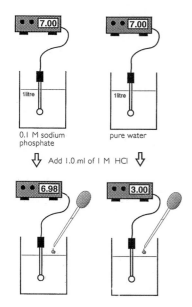

Figure 1.1 The concept of pH buffering

Add a few drops of strong acid to pure water or a solution of sodium phosphate, both at the same pH, and note the pH shift in both instances. If you are unsure about the derivation and use of the pH scale, don't worry—it will be discussed in some detail shortly.

buffers that could be used at each pH value—how do we choose which one to use? What factors must we consider before we can make this judgement?

In many instances, you will not need to worry about which buffer to use at all, as you will be following a published description, or more specifically, a recipe, that guides you in the preparation of the buffer. But there will be times when the description won't make sense, or where you will have to design a new buffer from scratch.

Whether you are following other people's instructions, or designing your own buffers, it is essential that you understand the principles of what you are doing. In this book we will cover the theory, and the practice, of buffer preparation.

2. Why do we need to worry about buffers?

You will know, of course, that the intracellular environment in most systems has a pH that is near to 7.0, i.e. neutrality—why should this be necessary? Both the hydrogen and hydroxide ions are really very reactive and can have many damaging effects on biological structures and processes. Quite small changes in pH can bring about profound disturbances in biological systems. Proteins contain many groups that can gain or lose hydrogen ions; when they do so they can acquire new structures that have quantitative or qualitative differences in biological activity. Many enzymes catalyse reactions that consume or generate hydrogen ions and the activity, as well as the structure of these enzymes is susceptible to pH changes. Changes in pH influence ion transport across cell membranes, energy production by the mitochondria (which depends on a gradient of hydrogen ions), and the light-dependent processes of chloroplasts.

Indeed, where we find big departures from the near-neutral 'physiological pH' it is usually for a specialist function: the lining of the stomach secretes 0.1 M hydrochloric acid (about pH 1) into the stomach to sterilize and denature ingested food, which increases digestibility. Lysosomes and endosomes maintain a low pH (about pH 4) inside themselves—perhaps to dissociate complexes or to denature proteins in anticipation of breakdown by the lysosomal proteolytic enzymes, the cathepsins.

◇ In fact, the concentration of H^+ doubles in acidosis from about 40 nM (pH 7.4) to 80 nM (pH 7.1) or halves in alkalosis, from about 40 nM to 25 nM (pH 7.6).

The pH of blood is usually about 7.4, and a relatively small shift in pH to 7.1 or 7.6 can elicit the clinical conditions of acidosis or alkalosis, both of which can require urgent medical treatment. These may sound like small shifts in pH, but we shall see later that a small change in pH reflects a much bigger change in the concentration of acid or alkali in a solution. It is, however, beyond the scope of this book to discuss the mechanisms whereby the body controls intracellular and extracellular pH.

Very few biological processes, whether *in vivo* or *in vitro*, are unaffected by pH changes—it is better to assume that control of pH should form an integral part of the design of **all** of your experiments.

3. Why do we need to know how buffers work?

◇ Some of you have never since done, and will never do another titration in your life!

You were probably introduced to 'pH and buffers' during your first year in further education. It is also a pretty safe bet that you didn't really enjoy this experience, and that you can remember very little about the details, except that you 'titrated some solutions'. Yet, in all the experiments that you will perform as a practising scientist, you will use buffers on a daily basis.

Some of these buffers will be provided in kits and some you will prepare using established recipes, but there will be buffer systems you have to devise for yourself. As a responsible scientist, you will try to understand what is going on in your experimental systems, and this understanding should extend to the behaviour of your buffers. But, if you want to become competent in the understanding, preparation, and design of buffer systems you will need to know a little background theory. It contains a little maths and some equations, but it is really very simple. This first chapter is a reminder of some of the basic concepts that you will need.

4. A word about 'equations'

We have heard the plaintive cry 'I'm no good at maths' too often to launch into the details of buffer theory without some explanation. At first glance you may be put off by the equations that you find scattered throughout this book. In our defence, we make the following pleas:

◇ This is a double-edged sword: we try to show you, step by step, how to calculate the answers—you see a lot of equations and formulae, rather than a one-line answer, and get the wrong impression about the complexity of the book!

We have explained each new equation in some detail as it is encountered. We expect some of you to be able to skip over the detailed explanations, because you will understand the concept already. However, if you are not clear about the ideas, work through the example with care—you will soon appreciate that this sort of maths is little more than straightforward arithmetic.

We have tried to illustrate new mathematic concepts with examples. Moreover, we have worked out these examples line by line, showing you how straightforward the calculations really are.

If you work through this book, we expect that you will be able to design and calculate most simple biological buffer systems for yourself. But we will also give you the opportunity to obtain from us, free, copies of software for MS-DOS/Windows® and Apple Macintosh® computers as well as World-Wide Web software that will allow you to design and produce the recipes for any buffer you are likely to want. The software is described in Chapter 6. As with most other software, our programs work on the 'garbage in–garbage out' principle. Having the software will not stop you making mistakes—a good awareness of the theory will.

So, please bear with us when we make mathematical diversions. You cannot really understand how buffers work without at least some understanding of the principles that underpin their action, and these principles are better described in equations than in words. We'll try to do both.

5. How to use this book

How you use this book will depend on your current level of understanding and your present needs. You will usually need to know how to use a pH meter—a piece of laboratory equipment that is subjected to constant abuse and misuse! Chapter 2 takes you through the basics of acid/base theory. Chapter 3 covers the theoretical principles of buffer behaviour; you may not need to read this chapter immediately, but at some time you should come back to it—the concepts are neither difficult nor esoteric. Chapter 4 gives a lot of background, and sound practical advice on the use and care of pH meters and electrodes. Chapter 5 discusses the practicalities of buffer preparations, and includes a series of 'case studies' that are worked examples of the design of buffers for particular purposes. Chapter 6 describes briefly the software that we have written to ease buffer design. Finally, the Appendices present detailed data on the most common buffers that are used in the biological sciences. At times, you will want to use the book as a data sourcebook, at other times as practical manual or a theoretical guide. It is not really necessary to read this book from start to end.

◇ Actually, when we understand and can design buffers properly, we could dispense with a pH meter completely, although few of us will have the courage to do this!

6. The range of proton concentrations

Implicit in your reading of this book is the need for a good understanding of the pH scale used to describe proton concentrations. If you are happy with the background to the pH notation and its usage, you can skip the remainder of this chapter.

The hydrogen ion concentrations that we find in biological systems are generally very low; typically in the range of 10^{-12} M (1 pM) to 10^{-1} M (0.1 M).

One litre of a 1 M solution contains 6.023×10^{23} molecules of a solute, we are therefore talking about solutions containing about 10^{11} (100 billion) hydrogen ions/litre to 10^{22} (10 000 billion billion) hydrogen ions/litre. We do not routinely describe our solutions in these terms, because it is clumsy and prone to error. Also, hydrogen ion concentrations have not traditionally been described in molarities. Rather, we conform with the widespread adoption of the pH notation devised by Sörensen. But there is nothing magical, nor is there a fundamental chemical 'truth' about the concept of pH—this notation was devised, and is still used, as a convenience.

◇ Avogadro's number, the number of molecules in a mole of material, is 6.023×10^{23}.

◇ Remember that:
US billion = 1 000 000 000 000 (10^{12})
UK billion = 1 000 000 000 (10^{9})

7. The pH scale

In science, as in nearly all other aspects of our lives, we use, predominantly, a number system based on 10. A consequence of this decadic system is that we talk a lot about 'log to the base 10' (abbreviated $\log_{10}$), mostly as a convenient way of expressing numbers that cover a large range. We can define x, the $\log_{10}$ of any number n, so that $10^x = n$. Thus, the $\log_{10}$ of 10 = 1 (because 10^1 is equal to 10), $\log_{10}$ of 100 = 2 (because 10^2 = 100), $\log_{10}$ of 1 000 000 = 6 (10^6 = 1 000 000), and so on.

◇ This is because we have ten fingers and thumbs—had we evolved three or five fingers and a thumb on each hand, the pH scale would have been based on $\log_8$ or $\log_{12}$.

Hydrogen ion concentration in biological systems covers a huge range, and a logarithmic notation would seem to be ideal. All of the hydrogen ion concentrations that we routinely encounter, however, are less than 1 M, and so, the $\log_{10}$ values of the molarities are all negative. $\log_{10}$ of 0.1 (10^{-1}) is -1, $\log_{10}$ of .000 000 1 (10^{-6}) is -6 and so on. So, just taking logarithms of the hydrogen ion concentration from 1 pM to 0.1 M would give us a scale from −12 to −1. To make the pH scale even more convenient, Sörensen made one final modification, and represented the pH scale as **the negative log** of the hydrogen ion concentration. Thus, the range described above runs from −(−1) to −(−12); i.e. from pH 1 to pH 12. In formal notation, we say that:

$$pH = -\log_{10}[H^+]$$

◇ In words: 'pH is minus the log base 10 of the hydrogen ion concentration'. You will probably know that we use square brackets to indicate the concentration of a species in solution.

Indeed, most biological processes, and most biological experiments are conducted in the range of hydrogen ion concentrations from 10^{-8} M (0.01 µM) to 10^{-6} M (1 µM). Thus, we are predominantly concerned with a pH range of 8 to 6 although we can have pH values greater than 14, and less than 1.

How does pH relate to acidity? A solution containing a very low concentration of hydrogen ions, say 10^{-11} M (10 pM), will have a pH of 11. An acidic solution might have a hydrogen ion concentration of 0.01 M, which is 10^{-2} M, or pH 2. Thus, the more acidic a solution, the lower the pH.

Because of the logarithmic nature of the pH scale, a change in pH of one unit is equivalent to a 10-fold increase or decrease in hydrogen ion concentrations. You will need to keep this in mind when you use buffers, because a small difference in pH value will mean a much bigger difference in hydrogen ion concentration. And, of course, it is the changes in hydrogen ion concentration, not the pH, that we really need to worry about.

8. Why is pH 7.0 'neutral'?

From where is the definition of pH 7.0, the pH of a 'neutral solution', derived? Consider a solution of pure water, which contains no added acid or alkali. Water will undergo a slight dissociation into the two ions, H^+ and OH^- in a reversible equilibrium. This equilibrium:

$$H_2O \rightleftharpoons H^+ + OH^-$$

is determined by an equilibrium constant, K, according to the simple expression:

$$K = \frac{[H^+][OH^-]}{[H_2O]}$$

which has a value of 1.8×10^{-16} M at 25°C.

◇ Why is the molarity of pure water 55.56 M? Water has a density of 1 kg/l, so one litre of water weighs 1000 g. Therefore, the molarity of pure water is 1000 divided by the molecular weight, 18.

We can simplify this equation a little, by assuming that in all the systems we will study there is a huge excess of water. Pure water has a concentration of 55.56 M. Using this value, which we will also assume to be constant, we can rearrange the equation to:

$$K \cdot [H_2O] = [H^+][OH^-]$$

This in turn simplifies to:

$$K_w = 1.8 \times 10^{-16} \times 55.56 = 1 \times 10^{-14} \text{ M}^2$$

This constant, K_w is called the ionic product of water and, simply put, says that the product of the concentration of hydrogen ions and the concentration of hydroxyl ions is equal to 10^{-14} M^2.

So, consider the situation where the solution is neither acidic, nor alkaline. In other words, there is no predominance of hydrogen ions or hydroxide ions. Because of the relationship defined by the ionic product, this means that $[H^+]$ must be equal to $[OH^-]$ and their product must be equal to $10^{-14} M^2$. Both $[H^+]$ and $[OH^-]$ must therefore be equal to 10^{-7} M. Remember that we define pH as $-\log_{10}[H^+]$, giving a pH of 7.0 for a neutral solution.

At pH 7.0, the sum of the concentrations of hydrogen ions and hydroxide ions is as low as it can be. Decrease the pH, and hydrogen ions will increase in concentration and hydroxide ions will decrease. Increase the pH and the concentration of hydrogen ions will decrease and that of hydroxide ions will increase.

◇ All of this reasoning only applies to aqueous solutions. Concepts of pH are very different if we use organic solvents mixed with water, for example.

9. The hydronium ion

Throughout this book we will discuss the hydrogen ion, or proton, as if it exists in solution. In reality, this is something of a simplification. Water is an extremely good solvent for ionic compounds (you know this already by intuition, even if you don't know the physical chemistry that underlies the observation). Because of the structure of the water molecule, it can interact with charged ions very effectively (in other words, 'dissolving' them). The stronger the ion, the more able it is to interact with water. The hydrogen ion, which has no electrons, is much smaller than any other ion and, therefore, has a very high charge density. So high a charge density, in fact, that it reacts with a water molecule in a near-permanent association. The free hydrogen ion does not exist in aqueous solution, but should be described as the hydronium ion, which has the formula H_3O^+.

◇ The hydronium ion is still a very ionic species, and probably interacts with three or four other water molecules itself.

The consequence of this is that we should really refer to the hydronium ion rather than the hydrogen ion. We would therefore have extra water molecules in the equation that do not influence the calculations, other than to be balanced out. Rather than introduce this complexity, we will refer to the hydrogen ion or proton throughout this book, and leave it to you, the reader, to remind yourself that '*the authors mean hydronium ion really*'.

2 Acids and bases

- Strong and weak acids/bases
- pH of solutions of strong acids and bases
- Why some acids are stronger than others

1. Introduction

Nearly all of this book is about the use of weak acids and bases as buffers. It may be useful to some readers, therefore, to run through a brief reminder of the definitions and properties of strong and weak acids and bases. Again, this is not essential information, but it is rewarding to have some idea, for example, why trichloroacetic acid is more hazardous than acetic acid, or why phosphoric acid is very corrosive, yet a phosphate solution forms the basis of many 'physiological' buffers. Readers who feel themselves familiar with these basic principles can skip over this chapter.

2. Definitions

We all use the terms 'acid' and 'base' (or, almost as comfortably, 'alkali' instead of base) with ease in the laboratory, and at some time in the past we may have been taught a more rigorous definition. We also know that some acids are 'strong' and others 'weak'. This awareness is often reinforced by the hazard labels on the bottles, but that does not really substitute for a degree of theoretical understanding to match this practical awareness!

◇ 'Acid' and 'alkali' are derived from Latin (*acidus*, sour; *acidum*, vinegar) and Arabic: (*al qaliy*, ashes of a plant).

The terms 'acid' and 'alkali' were first used in recognition of their properties in solution. Acids possessed a sour taste, and alkalis were able to neutralize or reverse the action of acids. Between 1880 and 1890, Arrhenius developed a theory to explain the action of acids and bases which was based on their dissociation into ions. One of the ions produced by an acid (such as HCl or CH_3COOH) was the hydrogen ion, H^+, and all bases (such as NaOH) were considered to produce a hydroxyl ion, OH^-. Although a major advance in our understanding of these compounds, the Arrhenius theory ran into difficulties when compounds such as Na_2CO_3 were considered—although they exhibited the properties of bases, they clearly could not ionize directly to yield a hydroxyl ion.

In the 1920s, the Arrhenius theory was independently developed by Lowry and Brönsted to a general and more powerful basis. The Lowry–Brönsted definition of an acid is a species that tends to donate or lose a hydrogen ion (proton). A Lowry–Brönsted base has a tendency to gain or accept a hydrogen ion. In the Lowry–Brönsted model, HCl donates a hydrogen ion to a water molecule (and is therefore an acid). The water molecule, conversely, functions as a base in this reaction:

$$HCl + H_2O \rightleftharpoons Cl^- + H_3O^+$$

This reaction is freely reversible, and thus, the Cl^- ion can accept the proton from H_3O^+—it follows that H_3O^+ must be an acid (obviously) and also that Cl^- must be a base (less obviously). The relationship between HCl and Cl^- is formalized by defining them as a 'conjugate acid–base pair'. The chloride ion is the conjugate base of HCl. Similarly, H_3O^+ and H_2O are a conjugate acid and base, respectively.

How does this Lowry–Brönsted definition help us to explain why sodium carbonate is a base? When sodium carbonate is dissolved in water, it will ionize fully into sodium ions and carbonate ions. The carbonate ion then reacts with a water molecule:

$$CO_3^{2-} + H_2O \rightleftharpoons HCO_3^- + OH^-$$

The carbonate ion acts as a base, in abstracting a proton from a water molecule to form a bicarbonate ion. Thus, the bicarbonate ion is the conjugate acid of the base, carbonate. The residual OH^-, of course, makes the solution alkaline, which satisfies our intuitive idea of a base. Thus, the Lowry–Brönsted theory of acids and bases provides a satisfactory basis to explain the behaviour of all acids and bases, not only those that contain H^+ and OH^-.

The acid-base concept could be extended further, to that proposed by Lewis. He defined an acid as any substance that can accept electrons and a base as any substance able to donate electrons. Since Lowry–Brönsted bases donate electrons to a hydrogen ion, a Lowry–Brönsted base is also a Lewis base. The Lewis concept extends our definitions of acids and bases to include reactions that do not involve hydrogen ions; this is important in many areas of chemistry. In a discussion of pH buffers, in which we are concerned with control and maintenance of hydrogen ion concentration, the Lowry–Brönsted definition will serve us well.

2.1 Strong and weak acids and bases

An acid is therefore defined as a substance that has a tendency to donate hydrogen ions to an acceptor molecule. As the word 'tendency' suggests, some acids are much more effective at this than others—we can therefore talk of 'strong' and 'weak' acids. But this is not a classification with just two categories. The ability to donate hydrogen ions varies over a continuous scale.

A moment's thought will also show that the strength of any acid cannot be considered in isolation—we need to know the nature of the base to which the hydrogen ion will be transferred. The transfer reaction will be influenced by the strength of both the acid and the base. Again, you will know this intuitively, and would not dream of mixing strong hydrochloric acid with a strong sodium hydroxide solution—the hydrogen ion transfer is so efficient that it generates dramatic quantities of heat. Perhaps less intuitively, you might also know that if you were to mix a solution of the same molarity of acetic acid with a solution of a weak base such as methylamine, the effect would be much less dramatic.

If we want to measure the strength of an acid, we must consider its ability to transfer a hydrogen ion to a standard acceptor base—the most obvious that we might use is water, according to the general reaction:

$$\text{conjugate acid} + H_2O \rightleftharpoons \text{conjugate base} + H_3O^+$$

The quantitative index of acid strength that we shall use is that defined by the dissociation constant. For the general reaction defined below, the equilibrium constant, K_{eq} for the reaction is simple:

$$HA + H_2O \rightleftharpoons A^- + H_3O^+$$

$$K_{eq} = \frac{[H_3O^+][A^-]}{[HA][H_2O]}$$

Because we are dealing mostly with dilute solutions of electrolytes, the concentration of water can be considered to be fixed, and we refine our definition of the constant K_a by taking into account this fixed concentration of water (see Chapter 1) at 55.5 M:

$$K_{eq} \cdot [H_2O] = \frac{[H_3O^+][A^-]}{[HA]}$$

and thus define an 'acid dissociation constant', $K_a = K_{eq} \cdot [H_2O]$. We now obtain the simpler form of the equation:

$$K_a = \frac{[H_3O^+][A^-]}{[HA]}$$

A moment's inspection of this equation will confirm that the larger the value of K_a, the more dissociation into A^- and H^+ there will be. Thus, a high value of K_a means a stronger acid (Table 2.1).

Table 2.1 Dissociation constants for a few acid/base pairs

The larger the value, the stronger the acid. Note that all of these constants are less than one. The units are molarity, as inspection of the equations above will confirm.

Acid	Conjugate acid/base	K_a(M)
Trichloroacetic acid	CCl_3COOH/CCl_3COO^-	2.2×10^{-1}
Phosphoric acid	$H_3PO_4/H_2PO_4^-$	7.5×10^{-3}
Chloroacetic acid	$CH_2ClCOOH/CH_2ClCOO^-$	1.4×10^{-3}
Formic acid	$HCOOH/HCOO^-$	1.8×10^{-4}
Acetic acid	CH_3COOH/CH_3COO^-	1.7×10^{-5}
Dihydrogen phosphate	$H_2PO_4^-/HPO_4^{2-}$	6.2×10^{-8}
Ammonium ion	NH_4^+/NH_3	5.6×10^{-10}

◇ In words, 'pK_a is minus the log (to the base 10) of the acid dissociation constant', or
$pK_a = -\log_{10} K_a$

Using the logarithmic notation described in Chapter 1, we can take the negative logarithm of the K_a value, which gives us, using the same 'p' notation as for pH, the familiar term pK_a. For the examples listed in Table 2.1, we can therefore calculate a pKa value (Table 2.2).

Acid	K_a(M)	pK_a
Trichloroacetic acid	2.2×10^{-1}	0.65
Phosphoric acid	7.5×10^{-3}	2.12
Chloroacetic acid	1.4×10^{-3}	2.85
Formic acid	1.8×10^{-4}	3.74
Acetic acid	1.8×10^{-5}	4.76
Dihydrogen phosphate	6.2×10^{-8}	7.20
Ammonium ion	5.6×10^{-10}	9.25

Table 2.2 The pK_a values for the same acid-base pairs as in Table 2.1

The stronger the acid, the lower the pK_a value (remember that the negative logarithm is used, which inverts the relationship between pK_a and strength, just as it does for pH and acidity).

A good understanding of the relationship between pK_a and K_a, and the relationship between K_a and the dissociation of an acid to yield protons, lets us understand the strength of an acid.

2.2 The role of water

Water can act as an acid or as a base, according to this simple relationship:

$$H_2O + H_2O \rightleftharpoons H_3O^+ + OH^-$$

This reaction proceeds to a small degree in pure water. If we adopt the convention adopted earlier, of ignoring changes in the water concentration because it is so large as to be essentially constant, we obtain a simple expression:

$$K_w = [H_3O^+][OH^-]$$

◇ For our purposes, pure water is defined as water containing no added neutral salts, acids or bases.

Under standard conditions, the value of K_w is $10^{-14} M^2$. Pure water is often described as neutral, having an excess of neither hydrogen ions nor hydroxyl ions (Chapter 1). From the definition of K_w, this gives us the result that in a neutral solution, the hydrogen ion concentration must be equal to the hydroxyl ion concentration. Both are at $10^{-7} M$, which yields a pH of 7.0.

◇ For example, acetic acid has a pK_a of 4.76 and ammonium chloride has a pK_a of 9.25.

Thus, water and pH 7.0 provide the reference point. Any substance with a pK_a less than 7.0 prefers to donate protons to water, and thus is acidic. By contrast, any substance with a pK_a greater than 7.0 will tend to act as a conjugate base relative to water, and abstract protons from water, leaving residual OH⁻ ions. These substances are therefore basic, yielding alkaline solutions.

3. The pH of a solution of a weak acid or base

When an acid is dissolved in water, it will dissociate, according to its strength, and we ought to be able to calculate the hydrogen ion concentration, and hence the pH. The general solution to this problem is straightforward, although it requires a little maths. Take a simple monoprotic acid, HA, such that:

$$HA + H_2O \rightleftharpoons A^- + H_3O^+$$

Intuition will tell us that the two parameters that affect the acidity of the solution of HA are first, the K_a value (usually expressed as pK_a, of course) and, secondly, the concentration of HA.

◇ Why can we assume that $[A^-] = [H_3O^+]$? Examine the equilibrium at the top of the page to remind yourself.

Two reasonable assumptions are made. First, a proportion of the acid, HA will dissociate to give A^- and H_3O^+, and this will therefore deplete the initial concentration of HA, defined here as C_i. Second, we can assume that $[A^-] = [H_3O^+]$. Thus, in the solution of the acid:

◇ C_i is the initial concentration of the acid.

$$[HA] = C_i - [H_3O^+]$$

We can use this relationship in the definition of the acid dissociation constant:

$$K_a = \frac{[H_3O^+][A^-]}{[HA]}$$

By substituting for both the [HA] term (which, remember from above, is equal to $C_i - [H_3O^+]$) and the $[A^-]$ term (which is equal to $[H_3O^+]$), we obtain an equation that relates C_i, K_a and H_3O^+:

$$K_a = \frac{[H_3O^+] \cdot [H_3O^+]}{C_i - [H_3O^+]}$$

The $[H_3O^+]$ is, of course, the term that gives us the pH of the solution, and we need therefore to solve the equation for H_3O^+. The equation is simply rearranged:

$$K_a \cdot (C_i - [H_3O^+]) = [H_3O^+]^2$$

which, with a little more work (expanding the brackets and moving terms to the left) gives us a recognisable form to the equation:

$$[H_3O^+]^2 + K_a \cdot [H_3O^+] - K_a C_i = 0$$

This equation is a quadratic, of the simple form $ax^2 + bx + c$, and thus, we can find the roots of the equation, using standard textbook mathematics.

$$[H_3O^+] = \frac{-K_a + \sqrt{K_a^2 + 4 \cdot K_a \cdot C_i}}{2}$$

Although there are two roots for this equation, we only ever use the positive '+' root, because we are discussing real solutions here (excuse the pun!), and the physical reality demands that $[H_3O^+]$ always be positive. An example will assist here:

Q: What is the pH of a 0.01 M solution of acetic acid?

◇ Remember that if $-\log_{10} a = b$, then $a = 10^{-b}$.

We have $C_i = 0.01$ M and, $K_a = 10^{-pK_a}$, $= 10^{-4.76}$, $= 1.8 \times 10^{-5}$

Substituting in the equation above, we obtain:

$$[H_3O^+] = \frac{-1.8 \times 10^{-5} + \sqrt{(1.8 \times 10^{-5})^2 + 4 \times 1.8 \times 10^{-5} \times 0.01}}{2}$$

$$[H_3O^+] = 4.16 \times 10^{-4} \text{ M}$$

$$pH = 3.38$$

The quadratic equation looks fairly complex and it is easy to make mistakes squaring and taking square roots of terms with negative exponents. If we look at the equation on the previous page again it may be possible to simplify the calculation somewhat:

$$K_a = \frac{[H_3O^+] \cdot [H_3O^+]}{C_i - [H_3O^+]}$$

If the $[H_3O^+]$ is substantially lower than C_i, then it is easy to see that the term $C_i - [H_3O^+]$ will not be very different from C_i. Thus, the equation simplifies to:

$$K_a = \frac{[H_3O^+]^2}{C_i}$$

which is rearranged to give us:

$$[H_3O^+] = \sqrt{K_a \cdot C_i}$$

This simplified equation can be made even more useful by converting it to logarithmic form, by taking $-\log_{10}$ of each side:

◇ In words, 'The pH of the solution is given by one half of the difference between the pK_a and the base ten log of the total acid concentration'.

$$pH = \frac{1}{2}(pK_a - \log_{10} C_i)$$

This simple form of the relationship holds true under two conditions. First, the concentration of the acid must be much greater than the hydrogen ion concentration. Secondly, the extent of ionisation of the acid must be relatively slight. As a rule of thumb, use the simplified formula when the concentration of acid is greater than 10mM, and the pK_a is between 3 and 11. Using the previous example, where C_i=0.01M, and pK_a=4.74, we now have:

$$pH = \frac{1}{2}(4.76-(-2))$$

$$pH = 3.38$$

For a solution of a weak base, with a pK_a between 7 and 11, the calculations are very similar. We have to adjust for the role of water as the hydrogen ion donor, and this is reflected in the introduction of K_w, the ionic product for water. A solution of weak base has a pH as defined by the approximation:

$$pH = \frac{1}{2}(-\log_{10} K_w + pK_a + \log_{10} C_i)$$

◇ $-\log_{10} K_w = -\log_{10} 10^{-14} = 14$.

$$pH = \frac{1}{2}(14 + pK_a + \log_{10} C_i)$$

Q: What is the pH of a 0.01 M solution of ammonium hydroxide?
The pK_a of ammonium hydroxide is 9.4 and, thus, the pH of the solution is given by:

$$pH = \frac{1}{2}(14 + 9.4 + (-2))$$

$$pH = \frac{1}{2}(21.4)$$

$$pH = 10.7$$

4. Structure and acidity

Thus far, we have accepted the premise that some acids and bases are 'stronger' than others. But why should this be? Why is hydrochloric acid stronger than acetic acid? Why is trichloracetic acid stronger than acetic acid? A treatise on factors that influence acidity (and conversely, basicity) is beyond the scope of this book, but a few general principles might illuminate.

The Lowry–Brönsted definition of acids means that we should consider the strength of the bond H–A that breaks to yield a hydrogen ion and the conjugate base. Compounds such as HCl are extremely polarized, and the strong electron-withdrawing property of the chlorine atom means that it tends to abstract the electron from the hydrogen atom. Thus, when HCl is dissolved, it separates almost completely into H^+ and Cl^-. The acid dissociation constant is very high, and HCl is a very strong acid.

With organic acids and bases, the situation is a little different. The ionizable group that yields a hydrogen ion may be less capable of releasing this proton, and thus is intrinsically weaker. Also, the ionizable group can be linked, through a series of carbon atoms, to a bewildering array of functional groups. How can these functional groups influence the acid dissociation constant (or pK_a) of the root ionizable group?

4.1 Ionizable groups in weak acids and bases

There are a large number of different functional groups that can ionize to yield protons, and which have pK_a values that are relevant in biological systems. Although it is not the purpose of this section to discuss these ionizable groups in detail, we can indicate the approximate acidity of each class (Table 2.3).

Table 2.3

The approximate pK_a ranges occupied by the types of functional group encountered in life sciences. The fact that a range of pK_a values is given for each functional group attests to the ability of surrounding influences to modify the acidity of the group.

By the way, even a functional group such as the C—H bond in alkanes has a pK_a of approximately 60. It is something of an understatement to say that this is not a very strong acid but, nonetheless, even this group can generate an occasional proton!

Acid	pK_a range
Monocarboxylic acids	3–5
Aliphatic dicarboxylic acids (1st dissociation)	1–4
α-Amino acids (carboxylate group)	2–3
Pyridinium ions	4–6
Aliphatic dicarboxylic acids (2nd dissociation)	5–7
Imidazolinium ions	6–7
Phenols	8–10
α-Amino acids (amino group)	9–11
Aliphatic aminium ions	9–11
Thiols	9–11
Guanidinium ions	11–14

4.2 Inductive and electrostatic effects

The ranges of the values for any functional group in Table 2.3 suggest that the local chemical environment of the functional group is important. Consider, for example, the series of aliphatic carboxylic acids (Table 2.4). They differ only marginally in pK_a values, which suggests that increasing the alkyl chain length has little influence on the ability of the carbonyl group to dissociate. Compare this lack of influence with the effect of chlorine atoms (Table 2.4).

Acid	K_a (M)	pK_a
Acetic acid (CH_3COOH)	1.8×10^{-5}	4.76
Propionic acid (CH_3CH_2COOH)	1.4×10^{-5}	4.86
Butyric acid ($CH_3(CH_2)_2COOH$)	1.5×10^{-5}	4.83
Acetic acid (CH_3COOH)	1.8×10^{-5}	4.76
Chloroacetic acid ($CClH_2COOH$)	1.4×10^{-3}	2.85
Dichloroacetic acid (CCl_2HCOOH)	4.5×10^{-2}	1.35
Trichloracetic acid (CCl_3COOH)	2.2×10^{-1}	0.65

Table 2.4

Two series of weak aliphatic acids, one modified by increasing alkyl chain length, the other by substitution of H by Cl. Note the lack of effect of alkyl chain, and the very strong effect of the chlorine atom.

Chlorine is very electronegative, and the strong electron withdrawing property of the chlorine atom is transmitted to the carboxyl group, making it less able to retain the proton. The proton is therefore released more easily, and the chlorinated acids are stronger. Two, or three chlorine atoms combine their effectiveness, and trichloracetic acid is really a rather strong acid, at a pK_a of less than unity.

◇ It might seem strange, at first glance, to have a pK_a less than one, but there is no reason why this should not be the case. Indeed, we can even have pK_a values that are zero or negative.

These inductive effects fall off rapidly as the two functional groups are separated. As you might expect, unsaturated bonds allow propagation of the inductive effect over a longer distance.

4.3 Mesomeric effects

Mesomeric effects arise when the ionizable group is stabilized by resonance of the electrons within the rest of the molecule. These particularly apply to aromatic groups, with additional functionalities in the *ortho-* or *para-* positions. Two examples will serve to illustrate these effects. First, the guanidinium group contains an imide nitrogen attached to the same carbon atom as the amino group. When this group acquires a proton, the delocalisation that the guanidinium group can undergo has the effect of stabilising that cation. Thus, the protonated form is stabilized. Conversely, loss of a hydrogen ion is therefore more difficult, and the guanidinium group is weaker than the amino group. Table 2.3 will confirm this expectation.

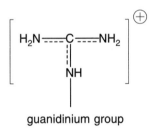

guanidinium group

By contrast, phenol is slightly basic, and yields a pK_a value of 9.99. Substitution of a nitro- group at the 4- (*para*) position makes the compound more acidic by over 2 pH units. Why should this be? The 4-nitrophenolate ion can undergo extensive resonance to generate several mesomeric forms. Thus, the phenolate ion (conjugate base) is stabilized by resonance, and the compound is a stronger acid (Figure 2.1). The pK_a value of 4-nitrophenol is 7.14.

Figure 2.1 4-Nitrophenol

The resonance isomers of the 4-nitrophenolate ion make it more stable, strengthening the acid.

4-nitrophenol

4-nitrophenolate ion

4.4 Steric effects

A double bond introduces restraints on rotation around a carbon–carbon bond, in the form of *cis-* and *trans-* isomers. If the isomers bring into proximity bulky, and especially charged groups, then they will interfere with each other's ionization. This effect manifests itself as a difference in pK_a values between the *cis-* and *trans-* forms of the molecule.

An excellent example is provided by fumaric and maleic acids. The differences between the pair of pK_a values for the two compounds are entirely due to the geometric isomers. The pK_a values for maleic acid are 1.91 and 6.33. For fumaric acid they are 3.10 and 4.60. In maleic acid, the second ionization is weaker, probably because of a combination of stabilization of the undissociated proton by the negative charge on the ionized carboxylate group, and hydrogen bonding between the undissociated proton and the carboxylate group.

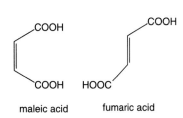

maleic acid

fumaric acid

4.5 Statistical effects

Consider the loss of a proton from a dicarboxylic acid in which each group has an equal probability of losing a proton. The acid can lose a proton in one of two ways (either group can ionize) but the proton can only be replaced in one way (since, ultimately, only one group did ionize). But, when the second proton dissociates, there is only one way to lose it, and two ways to replace it. This imbalance in probabilities yields the 'statistical effect', which has the outcome of modifying the pK_a values for the first and second ionizations. Consider butanedioic acid, with two identical carboxylic acid groups. Because of the ability to lose a proton in two ways, the first ionization constant, K_{a1}, is twice as large as would be expected for the closely related butanoic acid. Thus, the pK_{a1} is 0.3 ($\log_{10} 2$) units less than would be expected. Similar reasoning tells us that pK_{a2} will be 0.3 units greater than anticipated. For a compound with n identical groups, and n ways of losing the first proton, the pK_{a1} will always be less than expected by $\log_{10} n$.

butanedioic acid

4.6 Hydrogen bonding

Hydrogen bonding can have a dramatic effect in modification of the pK_a of an ionizable group. For example, the compound 2,6–dihydroxybenzoic acid is dramatically stronger than benzoic acid (pK_a = 4.2) because hydrogen bonding stabilizes the carboxylate group. The same effect is completely absent in 3,5–dihydroxybenzoic acid (Figure 2.2).

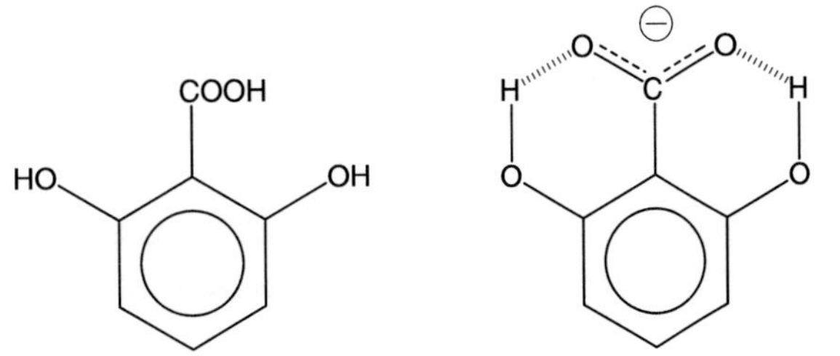

2,6–dihydroyxbenzoic acid (pK_a = 1.3)

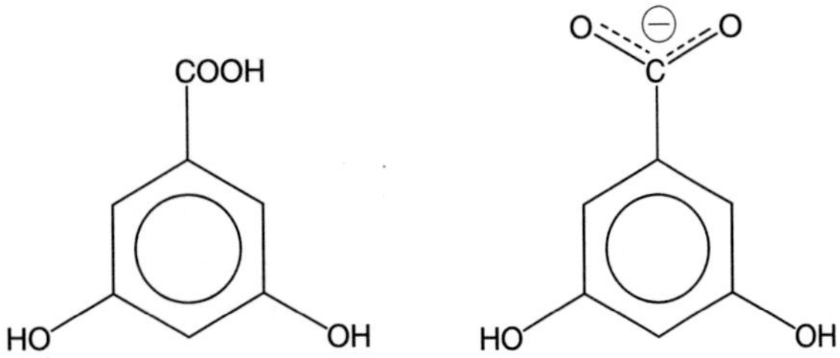

3,5–dihydroyxbenzoic acid (pK_a = 4.5)

Figure 2.2 Hydrogen bonding effects

The ability of the hydroxyl groups to hydrogen bond to the carboxylate group has a dramatic effect. The carboxylate group becomes much more stable, making this compound a much stronger acid.

Further reading

Perrin, D.D., Dempsey, B., and Serjeant, E.P. (1981) *pK$_a$ prediction for organic acids and bases*. Chapman & Hall, London.

Stewart, R. (1985) *The proton: applications to organic chemistry*. Academic Press, New York.

◇ Both books are rather advanced, and discuss the effects of structure on pK_a.

3 Theory of buffer action

- How buffers work
- Activity and concentration
- The Henderson–Hasselbalch equation
- Ionic strength effects
- Temperature effects

1. Introduction

In Chapter 1, we reviewed pH notation, and the importance of pH buffering, whether *in vivo* or *in vitro*. In this chapter, we need to cover some of the theory of buffer action, and introduce some new concepts. Some might argue that you could continue to prepare and use buffers without any of this theory. Nonetheless, an awareness of the important conclusions that derive from this theory will prevent you from making some of the more common mistakes.

A straightforward buffer, which 'buffers' pH changes, works in a very simple fashion. The objective is to maintain the concentration of hydrogen ions, H^+, at a fixed value, even if the process that we are studying is adding or removing H^+ from the solution.

How do buffers resist changes in $[H^+]$? To understand this process, we must remind ourselves of some simple chemical equilibria, in this case the equilibria of weak acids and bases. In the simplest form, we can write such an equilibrium as:

$$HA \rightleftharpoons H^+ + A^-$$

where HA is a weak acid, H^+ is the hydrogen ion (the proton), and A^- is the corresponding conjugate weak base. Three examples will help to familiarize you with these terms:

$$\text{acetic acid} \rightleftharpoons \text{acetate}^- + H^+$$

$$TrisH^+ \rightleftharpoons Tris + H^+$$

$$\text{phosphate}^{1-} \rightleftharpoons \text{phosphate}^{2-} + H^+$$

Acetic acid, phosphate^{1-}, and TrisH$^+$ are weak acids, and their conjugate bases are acetate, phosphate^{2-} and Tris respectively. Under physiological conditions, the species on both sides of the equation can co-exist in substantial amounts—compare this with a strong acid such as HCl, which is virtually completely ionized to H^+ and Cl^-. There are other more rigorous definitions of weak acids and bases which were alluded to in Chapter 2, but these need not concern us here.

◇ The structures of acetic acid and Tris are not really important here, as long as we appreciate that they can lose a hydrogen ion (Lowry–Brönsted acids). You will find their structures in Appendix 1.

◇ Nearly all pH buffers are weak acids or bases.

Notice that the weak acid can be neutral (acetic acid) or carry a positive (TrisH$^+$) or negative (phosphate^{1-}) charge. As we develop the theory of buffers, it will become clear that these charges on the buffer species have important consequences.

2. Weak acids and bases resist pH changes

A buffer is able to resist changes in pH because it exists in an equilibrium between a form that has a hydrogen ion bound (conjugate acid, protonated) and a form that has lost its hydrogen ion (conjugate base, deprotonated). For the simple example of acetic acid, the equation is:

$$CH_3COOH \rightleftharpoons CH_3COO^- + H^+$$

Here, the protonated form is acetic acid, with a net charge of zero, whereas the deprotonated form (acetate) has a charge of -1. The two species are in equilibrium, and this equilibrium, in common with all equilibria, can be displaced by addition of one component.

Consider a solution that contains equal amounts of acetic acid and acetate ions (10 mM acetic acid, 10 mM sodium acetate, for example). If we were to add a strong acid, such as HCl, to this solution, the added H$^+$ would displace the equilibrium to the left. Binding of H$^+$ to CH$_3$COO$^-$ 'mops up' the added protons (Figure 3.1). Electrical neutrality is preserved because every H$^+$ that reacts with a CH$_3$COO$^-$ anion to form the neutral CH$_3$COOH leaves behind a chloride (Cl$^-$) anion in its place. Add a strong base, such as sodium hydroxide, and the OH$^-$ ion would react with the H$^+$ and displace the equilibrium to the right. Electrical neutrality in the solution is sustained because for every CH$_3$COOH that is converted to CH$_3$COO$^-$, a corresponding Na$^+$ ion is added. This rule of electrical neutrality will be recalled later, when we calculate the overall ionic effect of buffers—there are always ions that are additional to those that buffer the pH changes.

This is the fundamental way in which buffers work. In this example, when we added a strong acid, the hydrogen ions did not all accumulate in the solution, but some of them became bound to acetate ions to form acetic acid. In other words, the buffer 'mopped up' some of the hydrogen ions, and thus resisted the change in pH that would otherwise have occurred. If we add a strong alkali, the OH$^-$ ions combine with the H$^+$ to form H$_2$O (remember this was discussed in Chapter 1) and, to compensate for this removal of protons, the equilibrium shifts to the right to release additional H$^+$. More acetic acid dissociates to release protons. The increase in pH that would have occurred because of the addition of alkali is thus minimized because the buffer compensates.

This is a simple enough concept, but to understand buffers properly, we have to develop the underlying theory a little more. It is time to introduce the Henderson–Hasselbalch equation.

As with all equilibria, we can express the dissociation of a weak acid or base in terms of an equilibrium constant. For the rest of this section, we will develop a general solution for an acid HA, but illustrate with two

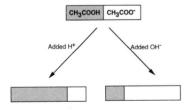

Figure 3.1 Buffer equilibria

A buffer solution in which both the conjugate acid and base are present will resist the addition of hydrogen or hydroxyl ions by shifting the equilibrium between the two species.

examples, acetic acid and Tris, two commonly used buffers. Consider once again the equation that introduces the equilibrium constant K_a.

$$K_a = \frac{[H^+][A^-]}{[HA]}$$

K_a is a constant, but is different for each buffer. For acetic acid/acetate it is $10^{-4.76}$ M, for Tris it is $10^{-8.1}$ M. What is the significance of these numbers, which seem to span a wide range, even judging by these two examples? To make things clearer, we shall modify this equation, using logarithmic transformation. First we rewrite the equation in a slightly different, but equivalent form:

◇ The corresponding equations for acetic acid and Tris buffers are:

$$K_a = [H^+]\frac{[CH_3COO^-]}{[CH_3COOH]}$$

$$K_a = [H^+]\frac{[Tris]}{[TrisH^+]}$$

$$K_a = [H^+]\frac{[A^-]}{[HA]}$$

Take $\log_{10}$ of each side of the equation, and remembering that when we have two terms multiplied together, we add their logarithms:

$$\log_{10}K_a = \log_{10}[H^+] + \log_{10}\frac{[A^-]}{[HA]}$$

Now, rearrange the equations to a different form, by swapping the two leftmost terms over, changing their signs as we do so:

$$-\log_{10}[H^+] = -\log_{10}K_a + \log_{10}\frac{[A^-]}{[HA]}$$

The left hand term should now be familiar: '$-\log_{10}[H^+]$' is of course, our definition of pH (Chapter 1). Similarly, we recognise an equivalent construction in the term '$-\log_{10}(K_a)$' and can refer to this as pK_a—we had already met this term briefly in Chapter 2. The equation now simplifies to:

◇ In words: 'The pH of a solution of a weak acid or base is given by the pK_a plus the log (base 10) of the ratio of the concentrations of base to acid.'

$$pH = pK_a + \log_{10}\frac{[A^-]}{[HA]}$$

This is the **Henderson–Hasselbalch** equation. In the general equation written above, the species HA is the acid (a proton donor) and the species A^- is the proton acceptor (a base). The general form written above is not particularly convenient, because it implies that the conjugate base always carries a negative charge, when what we really mean is that it has lost a proton relative to the conjugate acid, and thus, has decreased its charge by 1. Certainly this is true for a simple system like acetic acid, in which acetic acid is neutral and acetate has a charge of -1. For Tris, however, the base is neutral and, thus, when it picks up a hydrogen ion it must acquire a charge of $+1$. The acidic species in Tris has a charge of $+1$. It is therefore preferable to think of the Henderson–Hasselbalch equation in more general terms that will suit all cases:

$$pH = pK_a + \log_{10} \frac{[\text{base}]}{[\text{acid}]}$$

For acetic acid and Tris, the pK_a values are already defined by the value of K_a and, thus, we can substitute for the pK_a value in the equation. For acetic acid:

◇ This is why the Ka values were given in exponential form, as $10^{-4.76}$ M and $10^{-8.1}$ M, on the previous page. Knowing logarithms, it is easy to see that:

$$pH = 4.76 + \log_{10} \frac{[CH_3COO^-]}{[CH_3COOH]}$$

and for Tris:

$\log_{10}(10^{-4.76}) = -4.76$ and, therefore,
$-\log_{10}(10^{-4.76}) = 4.76$.

$$pH = 8.1 + \log_{10} \frac{[\text{Tris}]}{[\text{TrisH}^+]}$$

This relationship is of fundamental importance to an understanding of buffer action. It says, simply and clearly, that the pH of a solution is controlled by two factors; the relative concentrations of acid and base, and the pK_a of the conjugate acid/base pair. A few examples, using acetate and Tris, will clarify:

◇ We can ignore the counterion (Na⁺) because it does not contribute directly to the pH calculations (it has other effects, as we shall see later).

Q: What is the pH of a solution containing 0.01 M sodium acetate, 0.09 M acetic acid?

$$pH = 4.76 + \log_{10} \frac{[CH_3COO^-]}{[CH_3COOH]}$$

$$pH = 4.76 + \log_{10} \left(\frac{0.01}{0.09}\right)$$

$$pH = 4.76 + (-0.95)$$

$$pH = 3.81$$

Thus, under circumstances in which the concentration of the conjugate acid is greater than the conjugate base, we are satisfied that the system will equilibrate to release protons—the pH will therefore be less than the pK_a (but not necessarily acidic, defined as less than pH 7.0). Now consider the circumstances in which the concentration of conjugate base exceeds that of the acid.

Q: What is the pH of 0.09 M sodium acetate, 0.01 M acetic acid?

$$pH = 4.76 + \log_{10} \frac{[CH_3COO^-]}{[CH_3COOH]}$$

$$pH = 4.76 + \log_{10}\left(\frac{0.09}{0.01}\right)$$

$$pH = 4.76 + (+0.95)$$

$$pH = 5.71$$

And lastly, we should consider the special case where the concentration of conjugate acid and conjugate base are equal:

Q: What is the pH of a solution of 0.05 M acetate, 0.05 M acetic acid?

$$pH = 4.76 + \log_{10}\frac{[CH_3COO^-]}{[CH_3COOH]}$$

$$pH = 4.76 + \log_{10}\left(\frac{0.05}{0.05}\right)$$

$$pH = 4.76 + 0$$

$$pH = 4.76$$

◊ Remember that $\log_{10}$ of one is zero.

This gives us a fundamental property of all simple buffers—when the concentrations of acid and base are equal, the pH is equal to the pK_a. Any adjustment of the ratio of acid to base will cause the pH to shift away from the pK_a (Figure 3.2). Increase the amount of acid, and the term in brackets will become less than one, which means that it will have a negative logarithm and, thus, the pH is less than pK_a. Increase the amount of base and the term in brackets is greater than one, and will have a positive log, and therefore the pH will be greater than the pK_a.

Consider once again the role of the acid and base as 'source or sink' for hydrogen ions. If we could add acid or alkali to adjust the ratio of these two

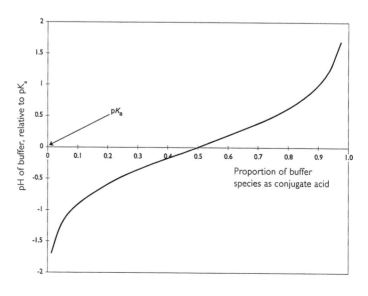

Figure 3.2

The response curve for pH as a function of the proportion of buffer that is the conjugate acid. As the proportion of buffer this is the conjugate acid moves towards 0.5, the pH approaches the pK_a value. Moreover, the slope of the line, which indicates how sensitive the pH is to changes in acid/base ratio, is at its lowest (the curve is most shallow at this point).

species, the pH would change. Look carefully at Figure 3.2 again, and estimate the pH change for a shift of 10% in the amount of acid or base (a shift of 0.1 on the *x*-axis of the graph). This shift, of 0.1, could, for example, be from 0.1–0.2, from 0.5–0.6, or from 0.9–1.0. Which of these three results in the smallest pH change? It should be obvious that the buffer will act best for hydrogen ions when the concentrations of acid and base are equal—the buffer will have maximal capacity to resist pH changes. Recall that when [acid] = [base], the pH is equal to the pK_a. Thus, a buffer works best at a pH range near to its pK_a. By contrast, if there was a very small amount of, for example, the acidic species, then the buffer would not be able to release many hydrogen ions to compensate for the addition of OH⁻. The further away the pH of a buffer is from its pK_a, the less effective it is as a buffer. One way to express this capability to resist changes is supplied by a graph of the slope of the line in Figure 3.2. This will change as the solution changes from predominantly acid to predominantly base (Figure 3.3). The lower the slope, the better the buffer is able to resist pH changes.

> The maximum ability to resist addition of a base occurs at the pK_a value.

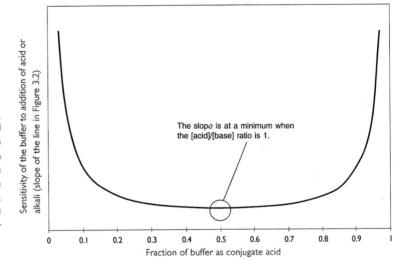

Figure 3.3

The slope of the curve in Figure 3.1 is plotted in this graph. If we were to zoom in on this curve, in the $x = 0.5$ region, we would find a minimum that is more closely defined. This curve only looks relatively flat because of the extreme range of *x* values that we have used. However, the inability of a buffer to resist pH changes at values (approx.) less than 10% or greater than 90% conjugate acid is clear.

A quick calculation will illustrate this point well.

Q: Two buffers are prepared by mixing acetic acid and sodium acetate. In both, the sum of acetic acid and acetate ions is 0.1 M. In the first, the [acetate⁻] is 0.01 M (Case A) and in the second, it is 0.05 M (Case B). Then, a strong solution of NaOH is added to a final concentration of 0.02 M. What pH changes do the two buffers undergo? (We shall ignore the volume expansion caused by addition of the NaOH solution.)

First, we have to calculate the pH of the buffers before the NaOH is added.

$$pH = 4.76 + \log_{10} \frac{[\text{acetate}^-]}{[\text{acetic acid}]}$$

◇ The [acetate⁻] defines the [acetic acid] of course. We already know the pK_a of acetic acid to be 4.76.

Case A: 0.01M acetate

pH = 4.76 + $\log_{10}$ (0.01/0.09)

pH = 4.76 + (− 0.95)

pH = 3.81

Case B: 0.05M acetate

pH = 4.76 + $\log_{10}$ (0.05/0.05)

pH = 4.76 + 0

pH = 4.76

Next, we add the NaOH to 0.02 M (conveniently ignoring the volume change for now!). The NaOH converts the equivalent amount of acetic to acetate ions:

pH = 4.76 + $\log_{10}$(0.03/0.07)

pH = 4.96 + (− 0.36)

pH = 4.40

pH = 4.76 + $\log_{10}$(0.07/0.03)

pH = 4.76 + 0.36

pH = 5.12

As expected, the pH rises in both cases. But, does the pH change by a different amount (calculated as ΔpH) in the two examples?

◇ We use ΔpH here to mean 'the change in pH', which, in this case, is caused by addition of the NaOH. And, lest you think that a ΔpH of 0.59 is quite a 'small' pH shift, you should realise that the change in [H⁺], from pH 3.81 to pH 4.4, is a decrease from approx. 160 µM to 40 µM—a four-fold reduction!

ΔpH = 4.40 − 3.81

ΔpH = 0.59 units

ΔpH = 5.12 − 4.76

ΔpH = 0.36 units

The buffer at pH 3.81 (Case A), with the greatest imbalance between [acid] and [base] shows the biggest pH shift. The difference is substantial—over 0.2 pH units between these two buffers. Thus, our expectations are confirmed—the more equally balanced the concentrations of the two species, the 'better' the buffer.

Note that the buffer in Case A is actually now better placed to resist a second addition of NaOH, because the two species are now more equally matched in concentration. As a general rule of thumb, this can guide buffer selection (Chapter 5) because if we have a system that generates protons, we should use a buffer with a pK_a slightly on the alkaline side of the pH we need. Then, as protons are generated, they shift the equilibrium to improve the balance between the buffer species, and the buffer becomes more resistant to proton additions. The converse arguments apply for a proton-consuming system, of course.

Thus, we should use buffers at or near their pK_a values. But intuition will tell us that the ability of a buffer system to resist changes must be influenced by the buffer concentration as well as the pK_a. For example, had we used a total of 0.2 M [acetate⁻] + [acetic acid] in the previous worked example, the pH changes (ΔpH) would have been approximately 0.35 (Case A) and 0.17 (Case B)—much smaller pH changes. Thus, three factors influence our choice of buffer. First is the pH that we need to maintain. Secondly, we should know whether our system generates or consumes protons. Thirdly, we should have some idea of the concentration of protons that are generated or consumed in our system. Already, the choice of a buffer is becoming more complicated!

3. How good is a buffer?—β values

We can define the ability of a buffer to resist pH changes in terms of 'buffering capacity' or β. This term measures how well the buffer works. The definition of β is inverted, because we express it as the amount of a strong base (defined as B, such as NaOH) that is needed to produce a change in pH of a fixed amount. In differential notation we refer to d[B]/dpH. The derivation of the equation for d[B]/dpH is beyond the scope of this book, and we shall adopt a simplified definition that works well between pH 3 and pH 11. This is given by:

$$\frac{d[B]}{dpH} = \beta = 2.303 \cdot \frac{K_a \cdot C_t \cdot [H^+]}{(K_a + [H^+])^2}$$

Where C_t is the total concentration of [acid] + [base]. This measure of buffer capacity therefore confirms our expectations that it depends on the value of K_a, the pH and the concentration of the buffer. Recall that the buffer should work best ($\beta = \beta_{max}$) when pH = pK_a. This is the same as saying $[H^+] = K_a$, and if we substitute those values (replacing both by x, for illustration), we get the following:

$$\frac{d[B]}{dpH} = \beta_{max} = 2.303 \cdot \frac{x \cdot C_t \cdot x}{(2x)^2} = 2.303 \cdot \frac{C_t \cdot x^2}{4x^2}$$

$$\beta_{max} = 2.303 \cdot \frac{C_t}{4}$$

◇ In practice, considerations or calculations of β are rarely made. Most systems in the life sciences do not generate or consume large numbers of protons and we tend to use relatively low concentrations of buffers so that other effects are minimised. Nonetheless, it is worthwhile considering whether your experiments do produce or consume protons in large amounts. Even in the worst cases, proton production should be less than 5–10% of the buffer concentration.

This confirms that even at the best buffering condition, pH = pK_a, the buffer capacity is still proportionally dependent on its concentration, but **only** on its concentration. This is intuitively acceptable as well—we are talking about the size of the proton sink or source.

4. pK_a is affected by the solution conditions

The equilibrium between the acidic and the basic species of any weak acid or base is affected by the solution in which the buffer is placed. This is more than just a theoretical problem, because the factors that influence pK_a, namely temperature and the concentration of ions that surround the buffer, both vary in our experiments, and therefore have a real practical significance. We must be aware of these influences on our buffers and how to correct for them.

4.1 Activity and concentration

When ions are in solution, they behave as though they are present at a lower concentration than expected. The activity is expressed as concentration multiplied by an activity coefficient, f, which must be less than or equal to 1. For the hydrogen ion, for example:

◇ The parentheses ( ) are used instead of brackets [ ] to indicate that we are considering activities rather than concentrations.

$$(H_3O^+) = f \cdot [H_3O^+]$$

Why should ions behave as though there are fewer of them than were initially dissolved? A detailed theoretical treatment is beyond the scope of this book, but a satisfactory explanation is that ions are shielded by the counterions of the opposite charge and, thus, act as if they are less 'active' than might be expected from their concentration. It follows that the more counterions there are, the greater the shielding, and this is exactly what is observed; the higher the concentration of ions, the greater the difference between activity and concentration (i.e., the lower the value of the activity coefficient). We shall see later that there is a relatively simple correction that allows us to modify our buffer design to accommodate this effect. This shielding can also occur when other ions are added to the solution—neutral salts, such as sodium chloride, can decrease the activity of a charged buffer species.

In practical terms, what does this mean? Two things. First, that if we add a neutral salt such as NaCl to a buffer, differential shielding of the conjugate acid and base will mean that the acid/base equilibrium will be affected, *and the* pH *will change*. Second, because the buffer ions shield themselves, simply *diluting a concentrated stock solution of a buffer will also cause the* pH *to change*. Fortunately, both complications can be substantially overcome.

Activity and concentration

At high concentrations, ions shield each other, reducing their influence. We say that their activity is lower than their concen-tration.

◇ Diluting stock solutions of buffers makes them change their pH. If you want to prepare a stock buffer, you must compensate for this change.

4.2 Ionic strength

A factor that is commonly overlooked in buffer design is ionic strength. The concept of ionic strength is used to describe the overall ionic properties of a solution, irrespective of the sign of the charge and chemical identity of the ionic species. Take a moment to clarify in your own mind a qualitative concept of 'ionic strength'—a measure of how 'ionic' the solution is.

It becomes obvious that a 100 mM NaCl solution has a higher ionic strength than 10 mM NaCl—there are 10 times as many ions (both Na^+ and Cl^-) in the solution at the higher concentration.

Similarly, a calcium ion (Ca^{2+}) can be considered as 'more ionic' than a Na^+ ion, because it has a higher charge density. Should a calcium ion be considered 'twice' as ionic as a sodium ion, or do we need to make a different correction?

We can now replace our qualitative impression of the 'ionic strength' of a solution with a rigorously defined description, the derivation of which need not concern us here. It looks intimidating at first glance, but is actually quite straightforward. The overall ionic nature of a solution is calculated as a parameter that is referred to as ionic strength, abbreviated I. Ionic strength is defined as:

◇ In words: 'Ionic strength is equal to one half of the sum of the concentrations of all ions multiplied by the square of their charges'.

$$I = \frac{1}{2}\sum_{i=1}^{n}(c_i \cdot z_i^2)$$

◇ The 'Σ' symbol (Greek, upper case sigma) means 'the sum of'. The numbers above and below the Σ sign tell us what we are adding to give the sum. We read this by looking at the bottom line first, which tells us that the term 'i' takes the value of 1 to n. In this example, n will refer to the number of ionic species that are present. Then, look at the symbols in parentheses after the Σ sign. The symbol 'c_i' refers to the concentration of the ith species, and the 'z_i' refers to the charge of the ith species—note that this is squared.

This may look daunting, and the symbols may be unfamiliar, but it is really just a shorthand notation for a very simple calculation. The ionic strength of a solution is dictated by each ion, and for each of the 'i' ionic species we must consider the charge, z_i and the concentration, c_i.

A couple of examples will help to make this clear. For a simple salt solution such as 0.1 M NaCl, we only have to consider two ions, the sodium ion that carries a charge of +1, and the chloride ion that carries a charge of –1. However, in the calculation, these charges are squared, and as such, the ionic strength term is always a positive number. Thus, for 0.1 M NaCl, the equation becomes:

◇ Although the equation may look more complex because of them, all of these different brackets, parentheses, and braces are used to show how the calculation is derived.

$$I = \frac{1}{2}\{([Na^+] \cdot +1^2) + ([Cl^-] \cdot -1^2)\}$$

$$I = \frac{1}{2}\{(0.1 \cdot +1^2) + (0.1 \cdot -1^2)\}$$

$$I = \frac{1}{2}\{0.1 + 0.1\}$$

$$I = 0.1$$

◇ Since the calculation of ionic strength is the sum of a series of ion concentrations, multiplied by their charges squared (a dimensionless number), ionic strength also has the units of concentration. However, it seems to be customary (though, pedantically speaking, incorrect) to refer to ionic strength as a dimensionless number—hence we would say that 0.1 M NaCl has an ionic strength of 0.1.

For a slightly more complex example, consider 0.1 M Na_2SO_4. In this example, we have 0.2 M sodium ions, and 0.1 M sulphate ions (which carry a charge of –2). So, the equation is more complicated:

$$I = \frac{1}{2}\{([Na^+] \cdot +1^2) + ([SO_4^{2-}] \cdot -1^2)\}$$

$$I = \frac{1}{2}\{(0.2 \cdot +1^2) + (0.1 \cdot -2^2)\}$$

$$I = \frac{1}{2}\{0.2 + 0.4\}$$

$$I = 0.3$$

In this example, the ionic strength is considerably higher than for 0.1 M NaCl, because of the strong effect of the doubly-charged sulphate ion, through the squaring of the charge. Solutions that contain multiply-charged ions will have the greatest potential to cause ionic strength effects.

Most enzymes and biological systems have evolved to work under 'physiological' conditions, which include a relative constancy of pH and of ionic strength. In mammalian cells and fluids, the ionic strength is approximately 0.154, which means that it has the same ionic strength as a solution that is 0.154 M NaCl.

◇ Such a concentration of sodium chloride is given by (0.154 × 58.5) g/l or 9.01 g/l. This is the same as 0.9 g/100 ml, which is why 0.9% NaCl is generally termed 'physiological saline'.

If we prepare a buffer solution in which the ionic strength is different from physiological, then we must not be surprised if the activity, solubility or stability of the components of the system change. Sometimes these changes can be useful, but unless we are aware of the behaviour of the ionic strength of the system, we are unlikely to be prepared for them.

4.3 Ionic strength depends on the pH of a buffer

Complications arise when we make up buffers, and forget (or, sometimes, conveniently ignore) ionic strength considerations. If we take a simple example, such as acetic acid/acetate, we already know that the relative amounts of the two buffer species will depend upon the pH of the solution. Thus, at pH = pK_a, the concentrations of acetic acid and acetate (as sodium acetate, for example—we must have a solution that is electrically neutral) are equal.

Acetic acid is unionized, and therefore does not contribute to the ionic strength calculation. The only ions that we need to consider are the acetate ion and the sodium ion. In a 0.1 M acetic acid/acetate buffer, at pH = pK_a = 4.76, we will have 0.05 M acetate and 0.05 M sodium ion. The calculation of ionic strength is simple, therefore, because both ions are monovalent, and thus the ionic strength has the same value as the total concentration of monovalent salt—in this example, $I = 0.05$.

What happens if the pH of the buffer is adjusted to pH = $pK_a - 1$ (i.e. 3.76)? In this instance, the Henderson–Hasselbalch equation tells us that there is 10 times as much acetic acid as acetate:

$$pH = 4.76 + \log_{10} \frac{[CH_3COO^-]}{[CH_3COOH]}$$

$$3.76 = 4.76 + \log_{10} \frac{[CH_3COO^-]}{[CH_3COOH]}$$

$$-1 = \log_{10} \frac{[CH_3COO^-]}{[CH_3COOH]}$$

$$0.1 = \frac{[CH_3COO^-]}{[CH_3COOH]}$$

Because there is 10 times as much acetic acid as acetate, the concentration of acetate is one-eleventh of the total = (0.1 × (1/11)) M = 0.009 M (or 9 mM) and thus, the ionic strength is also 0.009. At pH = $pK_a + 1$ (=5.76), there is 10 times as much acetate as acetic acid, and now, the acetate concentration is 10/11th of 0.1 M = 0.091 M (or 91 mM). Again, the ionic strength is the same as the concentration of acetate, and this time it works out to be 0.09 M.

Let's recapitulate. We have established that for a simple buffer such as 0.1 M acetic acid/acetate, the ionic strength will vary as the pH is changed. At pH 3.7, $I = 0.009$, at pH 4.7 it is 0.05 and at pH=5.7 it is 0.91. So, if you were to prepare three 0.1 M acetic acid/acetate buffers at the three pH values listed here, you would actually be making up three buffers at very different ionic strengths, covering a 10-fold range. This behaviour can also be expressed graphically (Figure 3.4).

This figure illustrates the dependency of ionic strength on pH of these buffers. For acetate buffers, the ionic strength increases as the pH increases, but for Tris buffers, ionic strength decreases as the pH is raised above the pK_a.

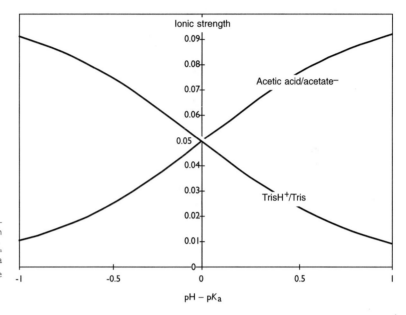

Figure 3.4

This plot indicates the change in ionic strength as we moved from pH = pK_a − 1 to pH = pK_a + 1 for two buffers—acetate and Tris, at a concentration of 0.1 M. Note two things: the direction and magnitude of the shifts.

If you wanted to use a series of acetate or Tris buffers to explore the influence of pH on a process, how would you be able to differentiate between effects caused by pH or ionic strength? There is no totally satisfactory solution to this problem, because if pH changes, the acid/base ratio must change. One approach is to prepare a series of buffers at constant ionic strength, and this is covered in Chapter 5. Ionic strength can be maintained by adjusting the concentration of buffer species, so that, for example, the acetate concentration is maintained at different pH values. However, the total buffer concentration will now change (another variable to deal with) as will the buffer capacity (which, recall from Section 3 of this Chapter, depends on the concentration of the buffer). An alternative approach is to make constant the total concentration of buffer species, and add extra neutral salts to sustain the ionic strength at a constant value.

4.4 Ionic strength influences the pK_a

We have established that the pH of a buffer (bearing in mind the nature of the ions and the pK_a values) influences ionic strength. But there is another, more subtle reason why would should worry about the ionic species in your buffers: *pK_a is also affected by ionic strength*! You will recall that earlier in this chapter we discussed, and then temporarily put aside, the fact that we should consider the *activity* of the buffer components rather than the concentrations. The more ions there are in solution, the more they 'shield' other ions, such as buffer components. A buffer will not function in quite the same way in the presence of dissolved salts as it does in their absence. If you make up a buffer solution, and then add salt to it, the pH, as well as the ionic strength, will change.

Although ionic strength will influence the activities of all of the ions, the effect on a buffer can be compressed into a single term— the effect of I on pK_a. The pK_a value will be modified if salt is added, or if the buffer species themselves contribute to ionic strength, and can be predicted from the following equation (sometimes known as the Debye–Hückel relationship):

◇ You may encounter alternative forms of this formula, that introduce the charges on the buffer species in a different way, or which include a second constant, B. All of these formulae, including the one shown here, are approximations, but all give similar answers in typical buffer calculations.

$$pK_a' = pK_a + (2z_a - 1) \cdot \left[\frac{A\sqrt{I}}{(1+\sqrt{I})} - 0.1 \cdot I \right]$$

In this equation, pK_a' is the modified pK_a value, z_a is the charge on the conjugate acid species, I is the ionic strength, and A is a constant that has a value of about 0.5, but which itself is temperature dependent (Table 3.1).

◇ The term pK_a' is sometimes referred to as a 'working' or 'practical' constant, and is sometimes cited at a typical ionic strength. See for example, the 'Good' buffers (Appendix 1).

To indicate the magnitude of the changes, the pK_a of acetic acid (where the charge on the acidic species, z_a, is 0) is *decreased* by 0.11 at an ionic strength of 0.1, and by 0.14 at I = 0.2. For a buffer such as Tris ($z_a = +1$) the pK_a is *increased* by the same amounts. These are quite large changes in pK_a (remember that because we are discussing changes in the logarithm of the number, they mean larger changes in the concentrations of hydrogen ions). Buffers that include species with charges greater than +1 or –1 show even greater effects, as the z_a term in the equation would imply. Thus, for the $H_2PO_4^{1-}$/HPO_4^{2-} buffer pair, the pK_a is decreased by 0.33 at an ionic strength of 0.1.

4.5 Dilution effects

Because ionic strength is dependent on the concentration of ions, it follows that buffers change their pH when diluted. If you prepare a stock buffer (say at 10 times working concentration) then when it is diluted the ionic strength will decrease by the same amount. Depending on the buffer, this will increase or decrease the pK_a'. The [acid]/[base] ratio is constant, and thus, from the Henderson–Hasselbalch equation, the only thing that can change when the pK_a' changes is pH.

If you make up a stock '10 X' (or 'ten-times concentrated') buffer solution, you should prepare it in such a way that when it is diluted it will yield the correct pH. Naturally, a 10 X buffer should only be used at 1 X; use it at any other dilution (such as 2 X or 5 X) and the pH, as well as ionic strength, will be different (see Chapter 5).

T(°C)	A
0	0.4918
10	0.4989
20	0.5070
25	0.5114
30	0.5161
37	0.5321
40	0.5262
50	0.5373
60	0.5494
70	0.5625
80	0.5767
90	0.5920
100	0.6086

Table 3.1 Values of constant A at different temperatures

4.6 Effect of temperature

The equilibrium constant of a buffer is a physical constant that obeys the laws of thermodynamics, which in practical terms, means that it is temperature dependent. This temperature dependence is most simply expressed as the term dpK_a/dT, which defines the change in pK_a with temperature. This parameter varies from buffer to buffer, and some are very temperature dependent—perhaps the worst culprit is also one of the most commonly used—Tris, with a dpK_a/dT of –0.028/°C. This means that for every degree change in temperature, the pK_a value will change by 0.028 units. Moreover, the sign on the temperature coefficient is negative, which tells us that as the buffer solution warms up, the pK_a value drops. An increase in temperature from 25°C to 37°C (12°C) will cause the pK_a to shift by $-(12 \times 0.028)$ units—from 8.1 to $(8.1 - 0.34) = 7.76$.

◇ The term 'dpK_a/dT' is also referred to as the temperature coefficient and derives from calculus notation meaning the rate of change of pK_a with temperature.

What are the practical significances of this temperature effect? We might encounter the effect under two circumstances—when the temperature of our experiment must change as it is being performed, and when we prepare a buffer at one temperature and then use it at a different temperature.

To illustrate, we will take the situation where we have a Tris buffer which we make up at 25°C to be pH 8.1—because the pH is nominally equal to the pK_a, we are adjusting the concentration of acidic and basic species to be equal. Now, let's change the temperature to 37°C, so that the pK_a shifts from 8.1 to 7.76. When the buffer is warmed up, the pK_a will change to 7.76, and we must substitute the new value in the Henderson–Hasselbalch equation. Between pH 3 and pH 11, the change in proton concentration will usually be small compared to the concentrations of buffer species, and the pH shift is, to a first approximation, equal to the change in the pK_a.

◇ Be assured that there will be instances where scientists have made up a Tris buffer at room temperature, and then used it at a higher (such as a 37°C incubator) or lower (e.g. a cold room) temperatures. Some of them will have forgotten or will have ignored the necessary correction for the temperature shift.

In other words, increasing the temperature of a Tris buffer makes the pH decrease, because the pK_a also decreases. A drop of 0.34 pH units from 8.1 to 7.76 corresponds to a change in hydrogen ion concentration from 7.9 nM to 17.4 nM. Although the pH change was 'only' 0.34 units, the hydrogen ion concentration has more than doubled as a consequence of the temperature jump! How many other ions in the buffer would one be this casual about?

A nice example of temperature effects is given by the polymerase chain reaction (PCR), which uses thermal cycling to amplify DNA by repeated cycles of replication *in vitro*. Many recipes for PCR use Tris buffers. As we have seen, temperature has a big effect on Tris, and it is instructive to calculate the pH change that the PCR reaction undergoes in a typical thermal cycle. The calculations used here are approximate, because most buffer theories are not very robust at high temperatures. The changes are shown in the diagram alongside (Figure 3.5). Note that we are not criticising the way in which PCR is done (it is a stunningly successful technique), but it provides a nice illustration of temperature effects. Incidentally, we note that a recent development in PCR has substituted other buffers for Tris—we wonder whether part of the improvement is attributable to the lower temperature coefficient that reduces the pH shift during the reaction.

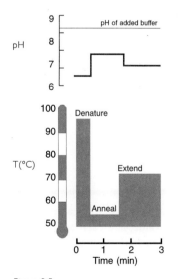

Figure 3.5

The effect of temperature cycling in PCR on the pH of the reaction mixture. The [H⁺] changes cyclically over a ten-fold range!

Not all buffers are affected by temperature in the same way. Amine buffers, such as Tris and triethanolamine are most affected, whereas anionic buffers such as phosphate, or simple organic acids such as acetic acid are much less affected. Appendix 1 lists the commonly used buffers together with their dpK_a/dT values—consult the reading list if you need other values.

5. Thermodynamic and apparent pK_a values

At this stage, we have made our simple buffer calculation much more complex. We now know the importance of ionic strength and temperature and, that the pK_a that we get from text books is not the one that operates in our solutions—it is modified by both temperature and ionic strength. We must calculate the working value pK_a' or the 'apparent pK_a' from the true, or 'thermodynamic' pK_a value. Fortunately, it does not matter which order we apply the corrections, because they are additive:

$$pK_a' = pK_a + I \text{ correction} + T \text{ correction}$$

The T and I corrections can, of course, be positive, negative, or, in some circumstances, zero. Whatever their magnitude, you should always include them in the calculation of the working pK_a.

The corrections assume that we know the ionic strength and temperature at which we will use the buffer. Temperature is usually easy, but when it comes to ionic strength correction, there is one final complication. If we wish to make a buffer of fixed final ionic strength, we can, knowing that ionic strength, calculate pK_a'. What do we do, however, if we want to prepare a buffer that is, for example, 0.1 M buffer species, but with no initial adjustment of ionic strength? To calculate the buffer composition properly, we will need to know the ionic strength. This is controlled by the concentrations of acid and basic species, and their counterions, which we get from the Henderson–Hasselbalch equation—for which we need to know the pK_a'. We cannot calculate pK_a' however, because we do not know the ionic strength! How can we get around this cyclical argument? This is covered in the Chapter 5 and the software described in Chapter 6 will perform calculations to circumvent this problem.

6. Polyprotic buffers

Polyprotic buffers have more than one ionisable group that can generate protons. Each group will have its own pK_a' value, and each will act as a buffer within about 1 pH unit on either side of each pK_a' value. How should we treat such polyprotic buffers in our calculations?

If the pK_a' values are sufficiently far apart, we can ignore all ionisations other than the one in which we interested. For example, phosphate buffers exist in four forms, from uncharged phosphoric acid to full deprotonated phosphate^{3-}. The pK_a' values for the three equilibria are so far apart (about 5 units) that to all intents and purposes, we can ignore the species that do

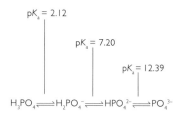

The four ionisation states of phosphoric acid

The pK_a values for the three ionisations are identified at the appropriate equilibrium.

not flank the equilibrium in which we are interested. For example, in a phosphate buffer at pH 7.00, the concentration of phosphoric acid or phosphate^{3-} are both so low that we can ignore them in our calculations of the buffer properties. The only species we need to consider are phosphate^{1-} and phosphate^{2-}. Thus, the calculations are no different from those for monoprotic buffers, as long as we recognise that some buffer species can have multiple charges, which because of the charge-squaring term, will have a relatively large effect on ionic strength.

◇ In general, it should not be necessary for you to use polyprotic buffers with overlapping pK_a values. There is a good selection of monoprotic buffers or buffers with far-separated pK_a values to choose from.

Other buffers, such as those based on succinic acid cannot be treated in this fashion, because the pK_a' values (4.21 and 5.64 in succinic acid) are sufficiently close that the equilibria overlap. The equations to calculate the concentration of each species are therefore more complex as well, and will not be derived here, but the following equations describe the composition of a dibasic buffer such as succinate calculated to a known ionic strength (I):

$$[A_t] = I \cdot \frac{\left(1 + \frac{[H^+]}{K_1'} + \frac{[H^+]}{K_2'}\right)}{\left(1 + 3 \cdot \frac{K_2'}{[H^+]}\right)}$$

$$[base] = I \cdot \frac{\left(1 + 2 \cdot \frac{K_2'}{[H^+]}\right)}{\left(1 + 3 \cdot \frac{K_2'}{[H^+]}\right)}$$

$[A_T]$ is the total concentration of the acid buffer species and [base] is the concentration of alkali which needs to be added to it to give the correct pH and ionic strength. K_1' and K_2' refer to the two ionization constants, appropriately corrected for temperature and ionic strength.

7. 'Mixed' buffers

◇ We do not advocate the use of these mixed buffer systems. In our experience it is preferable to use a range of buffers containing a single buffer species, and adjust the ionic strength so that many of the buffer-specific ionic effects are diminished.

You will encounter 'odd' buffers that seem to comprise two different buffering species. An example commonly encountered is that of Tris/acetate, in which two buffering components are present. The logic that underpins such buffer design may be hard to discern. Tris will buffer in the region of pH 7 to 9, and acetate from pH 4 to 6. At the higher pH range, effectively all of the acetic acid is present as a counterion, acetate$^-$ and, thus, is functioning in the same way as the chloride ion in a simple TrisHCl buffer. At the lower pH range, all of the Tris is present as the TrisH$^+$ ion and thus is acting in a similar fashion to the sodium ion in a simple acetic acid/sodium acetate buffer. Why use one buffer species as a counterion for a second?

The historical basis of such buffer designs stems from an era when buffers were limited to a few readily available components at adequate purity. Tris/acetate would provide good buffering capability over nearly all of the pH range from 4 to 9 (albeit with an area of dubious buffer capacity between pH 6 and 7) and thus could be used as a broad-range buffer to explore biological systems.

When looking at the effect of pH on many biochemical processes it may be necessary to have the same buffer components present over the whole range of pH values to be studied. Only then could any effects be attributed to variation of pH rather than to change of buffer components. However, it is reasonable to surmise that only the conjugate acid or conjugate base is having an effect and, thus, the argument falls down somewhat, because the concentration of these species will vary as the pH changes. The only way in which an acetate buffer can be prepared at a constant concentration of acetate⁻ ion over several pH values is by modification of the total buffer concentration. As the pH rises, so less buffer is needed to create the required acetate⁻ concentration, because the equilibrium is shifting in favour of the anionic species.

◇ If you suspect buffer-specific effects, you should repeat the experiment using a buffer system that is chemically very different (in charge and overall polarity, for instance), again controlling for ionic strength.

8. Buffers in heavy water

All that we have so far considered has applied to simple aqueous solutions. Sometimes it is necessary to prepare buffers to work in heavy water (deuterium oxide, D_2O) particularly for NMR, or where where isotopic labelling experiments and studies of reaction mechanism are involved. It is beyond the scope of this book to go into details—if you need to do this type of experiment you will be aware of the problems. The measure of acidity in heavy water as the solvent is known as pD and may be determined in the same way as pH (using a pH meter with a glass electrode). If the pH meter is standardized against aqueous buffers however, the pH meter reading must be corrected when heavy water is present. The appropriate relationship is:

$$pD = pH_{(meter\ reading)} + 0.4$$

Further reading

◇ These books contain more formal discussions of theory and useful sets of buffer tables. The last includes a discussion of buffers for control of metal ion concentrations as well as of the hydrogen ion.

Bates, R.G. (1973) *Determination of pH: theory and practice.* Wiley, New York.

Dawson, R.M.C., Elliot, D.C., Elliot, W.H. & Jones, K.M. (1986) *Data for biochemical research.* Clarendon Press, Oxford.

Perrin, D.D. and Dempsey, B. (1974) *Buffers for pH and metal ion control.* Chapman & Hall, London.

4 Measuring pH

- The pH meter
- pH electrodes
- Calibration of the meter
- Care of electrodes

1. Introduction

Having defined pH, we must address the thorny problem of measuring the pH of a solution. This requires a pH meter—maybe the most-abused item of equipment in a typical laboratory! The pH meter consists of two parts. First, there is an electrode system which, when dipped into the solution of interest, generates a voltage in a pH-dependent fashion. The electrode is usually connected to a voltmeter which can measure the voltage, convert that voltage into the corresponding pH value, and display the result. Sometimes, probe type meters have the electrode and voltmeter integrated into the same package—such are the miracles of miniaturization, semiconductors, and low power devices.

2. The electrode

The electrode must respond to the concentration of hydrogen ions in solution. The electrode system beloved of physical chemists has historically been the hydrogen electrode; this consists of gaseous hydrogen bubbled over a platinum surface placed in the solution and is totally impractical for general laboratory use! Luckily the glass electrode is a good substitute for the hydrogen electrode.

2.1 The glass electrode

Let us start by considering the business-end of our measuring system, namely the glass electrode. This consists of a very thin, and therefore fragile, glass bulb which contains a dilute acid solution and a connecting wire. When the glass bulb is placed in the solution under test, protons from the solution stick to the surface of the glass and generate a voltage across the glass membrane. It is fair to say that the chemistry of this process is still not fully understood. However, this need not worry us as we are merely interested in the electrode as a device for generating a response to pH. The voltage generated at the glass electrode can only be measured by reference to another electrode system and that usually employed is either the Ag/AgCl electrode or the calomel electrode. The reference electrode also needs to be placed in the test solution.

◇ We will not venture into a detailed discussion of the electrochemistry of cells here. It is enough to know that the glass electrode generates a voltage when the concentration of protons is different on each side, and that other cells (such as calomel or Ag/AgCl) can generate a stable reference against which this voltage can be measured.

At this point there is a divergence of practice in pH measurement. Sometimes separate glass and reference electrodes are used, necessarily resulting in a clumsy system, but more commonly, 'combined electrode' systems are used. The combined electrode is conveniently compact, comprising a concentric design in which the glass electrode elements are surrounded by a KCl solution linked to the test solution by a semi-porous plug and also to the reference electrode. The latter therefore forms an integral part of the glass electrode system. So common and convenient is this concentric electrode design that most people refer to it as a glass electrode, unaware that they are actually using a dual electrode system. A typical glass electrode design is shown in Figure 4.1.

◇ The solution inside the electrode leaks very slowly, through the porous plug, into the test or soak solutions. This is why it needs to be topped up regularly.

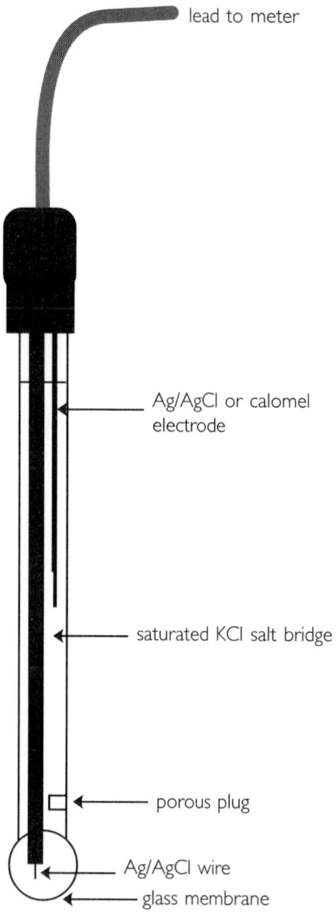

3. Care of a pH electrode

3.1 Keep the electrode filled

To maintain the correct function of the electrode, it is necessary to ensure that the potassium chloride solution (KCl bridge) is kept saturated and topped up. For this purpose, most electrodes possess a top-up hole at the top (usually stoppered in some way) and you should periodically add saturated KCl solution through this, as well as a few KCl crystals. KCl crystals should always be visible inside the electrode bridge. This helps to prevent the development of a large liquid-junction potential at the porous plug which will lead to erroneous pH measurements.

3.2 Mechanical abuse

The glass membrane is easily damaged and is often protected by a short sleeve of tougher glass or plastic. This protected design is worthwhile, as the electrodes are expensive and, if broken in your solution, will cause both expense and contamination of the solution.

3.3 Contamination of the electrode

With age, the electrode can become contaminated by lipid or protein present in biological test liquids. This sticks to the surface of the glass bulb and is usually noticeable as a sluggish response of the electrode system. That is to say, the electrode takes a long time (tens of seconds) to settle to a stable pH reading when placed in a test solution. Lipids may usually be removed by washing the electrode with acetone. Proteins are removed by overnight soaking in 0.1% pepsin in 0.1 M HCl. Pepsin is a proteolytic enzyme secreted into the stomach, and which shows optimal activity at the HCl concentration of the stomach secretions. This is why it is used at such a low pH! When cleaning the electrode, never be tempted to dry it by rubbing with a dry tissue. This may damage the electrode and will also generate static charges on the glass bulb which will affect the pH reading.

Figure 4.1

The combination glass electrode

(labels: lead to meter; Ag/AgCl or calomel electrode; saturated KCl salt bridge; porous plug; Ag/AgCl wire; glass membrane)

◇ Pepsin is rapidly and irreversible inactivated at pH 7.5, so don't be tempted to make a stock solution at neutral pH.

4. Replacing an electrode

◇ When should an electrode be replaced? Specific guidelines are difficult to provide, but you should embark on a troubleshooting exercise if any of the following happen:

- no response to different pH solutions,
- variation in pH reading with speed of stirring,
- a 'slope' or 'pH' setting on the meter that is much different to normal,
- a pH reading that changes when you move your hand towards the electrode,
- a slow, incessant drift in the pH reading obtained from a solution or standard.

Typically, a combination glass electrode should last several years if properly cared for. Until the advent of tougher and protected electrodes, the fragility of the glass bulb provided for a regular programme of electrode replacement!

◇ Keeping electrodes moist at a low pH is ideal—this ensures that the glass is properly hydrated and also that H^+ exchange for other ions that might be adherent to the electrode surface. Never store electrodes in distilled or deionized water—the low ionic strength of these buffers encourages the filling solution to leave the electrode.

Glass electrodes from different manufacturers are all basically the same and interchangeable. You do not have to buy your electrodes from the company which made the pH meter. On the other hand, many manufacturers put unusual plugs on their electrodes and strange sockets on their meters to discourage you from mixing electrodes and meters from different suppliers! If you are competent in electronics, you can remove the plug from the electrode lead and replace it with an appropriate substitute. Bear in mind that the plug is connected to the electrode by a screened, coaxial lead (similar to that attached to television and radio aerials) and it is essential to maintain the screening in order to obtain stable, drift-free pH measurements.

New electrodes should be soaked in 0.1M HCl for about 4 hours, followed by about half an hour in distilled water, prior to use. Electrodes should be stored, for short periods (day to day), in buffers between pH 4 and pH 7 or a dilute (0.05 M) HCl solution. For longer periods (weeks or months when they will not be used) they should be stored dry. The salt bridge of combined electrodes should be kept saturated, with KCl crystals visible, and the glass bulb should be gently cleaned with moist tissue during measurements, never with dry tissues. Glass electrodes also have a tendency to age. This results in a slow drift in the voltage generated for a given pH value. It is possible to re-calibrate the meter to compensate for this, but you should expect pH electrodes to be replaced regularly.

It is sometimes necessary to buy more expensive variants of electrodes (made of special types of glass) for specialist use such as determination of extreme pH values. In general, these can be treated in the same way as normal glass electrodes, but follow manufacturer's guidelines.

5. The pH meter

The pH meter measures a voltage that is generated by the electrode. The voltage (E, electromotive force, emf) generated by the glass electrode is given by the expression:

$$E = -\frac{2.303 \cdot R \cdot T}{F} \cdot \text{pH}$$

where R is the universal gas constant, T the absolute temperature and F is Faraday's constant. This is about –59 millivolts (mV) per pH unit at 25°C. This is a substantial voltage but, owing to the high resistance of the glass membrane, only gives a tiny current, which complicates measurement of the voltage. Traditional pH meters have been very high input impedance, direct current measuring devices based on valves (vacuum tubes) which made them relatively expensive. About ten or more years ago, analogue to digital voltage converters and liquid crystal display drivers became available in integrated circuit form. These devices, both compact and cheap, revolutionized pH meter design.

◇ Home-made meters have been used on a daily basis in our teaching laboratories for at least ten years. They are battery driven and fairly portable and have only required one or two battery replacements during this time. The original design of meter was very cheap — the glass electrode cost more than the rest of the meter!

◇ You do not have to worry about all the electronic terminology, but those with a technical bent will appreciate the details.

The current range of cheap meters on the market are accurate and easy to use. Of course, you can still spend a lot of money on a meter and this usually gives you added features controlled by a custom microprocessor. This generally makes the meters more complicated to use and can insulate the user from the knowledge of what they are doing. If you use one of the more sophisticated meters, you must take the time to understand it fully.

The basic design of a digital pH meter is shown in Figure 4.3. The high impedance glass electrode usually delivers its voltage into an amplifier that amplifies the current (but not the voltage) and which simplifies connections between the other components. The voltage generated by the electrode (which varies according to the pH of the solution) is compared with a reference voltage (generated in the meter). This voltage is then converted to a digital form to drive an alphanumeric display (which has replaced an analogue meter).

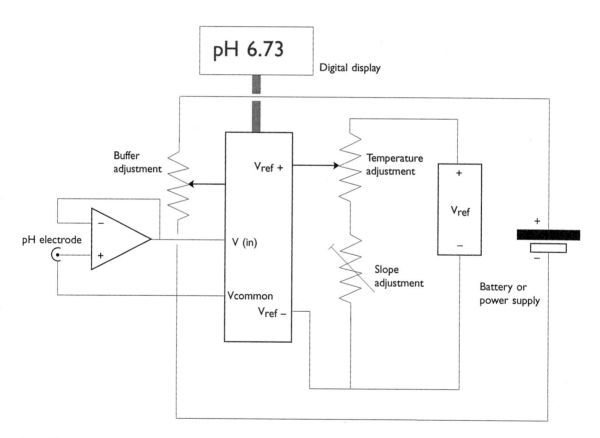

Figure 4.3

A schematic of a modern pH meter. This is not meant to be electrically correct, but indicates that the 'Buffer' control modifies the voltage from the pH electrode, and the 'Slope' and 'Temperature' controls affect the reference voltage. The 'Slope' control should only need adjustment at infrequent intervals, as the electrode ages.

6. Calibration of a pH meter and electrode

When we measure pH we are really measuring the *difference* in pH between our test solution and some form of pH standard. The voltage generated by the electrode depends on temperature, age and state of the electrode and is rarely equal to the theoretical value of 59mV/pH unit. For this reason we calibrate the meter against carefully prepared standards of known pH value. These solutions may be made from commercially obtained buffer 'tablets' or by calculation and preparation of appropriate buffer solutions. Some pH standards are listed in Appendix 2.

6.1 Temperature effects

◇ Although the voltage generated by the glass electrode is –59 mV/pH unit at 25°C, at 100°C it is –74 mV/pH unit and, at 0°C, it is –54 mV/pH unit.

The pH reading delivered by an electrode/meter combination will vary with temperature for two reasons. First, you will have noticed that the voltage generated by the electrode varies with temperature (the term T in the equation). This temperature must be made known to the pH meter before it can convert the voltage measured to the correct pH value. Often, a knob is provided on the pH meter labelled 'temperature compensation' or something similar; sometimes a probe is placed in the solution to measure the temperature and report it back to the meter.

The second influence of temperature is on the pH of the standard solutions and test solution. This dependence is a chemical process, due to the change in ionization of the buffer components with temperature (Chapter 3). You may be able to measure the pH of your test solution with great precision and accuracy at one temperature, but if you use it at a different temperature the pH will nearly always be different because the pK_a has changed at the new temperature.

◇ This makes sense even if you know what you are doing and is imperative if you do not!

Ideally, you should always adjust the temperature of your standard buffers to the temperature at which you wish to make the pH measurement. Ensure that you know the pH of the standard buffer at that temperature (not just at a single temperature specified by the supplier). It is best to use a standard buffer with a very small temperature dependence. Measure and use your test solutions at the temperature at which the pH meter is calibrated.

6.2 Calibration of a pH meter

◇ Expensive pH meters with inbuilt microprocessors may partially automate the calibration step for you.

Having controlled for temperature effects, we must standardise the pH meter. A pH meter, of course, measures the pH of our test solution by reference to a standard buffer. The electrode is first placed in the standard solution of known pH and the buffer adjustment control is operated until the meter reading corresponds with the known pH of the standard solution at the selected temperature. Some standards are listed in Appendix 2. Errors in the pH measurement will be small if the standard and test solutions are of similar pH.

6.3 Slope adjustment

A second standard buffer is routinely used to calibrate the meter. You will have noticed (Section 5) that the voltage generated by a glass electrode is a linear function of pH. But, the slope of this line may not be -59 mV/pH unit, and a second buffer can be used to compensate for this variation. First, standardize the meter against the first buffer using the buffer adjustment knob. Next, place the electrode in the second buffer and see if it gives the correct reading. If it does not, adjust the slope control until it does. This process may need to be repeated in a cyclical manner until the readings are correct for both standards.

◇ This slope adjustment should be made regularly, perhaps once a week, to ensure proper function of the meter.

◇ For a single point calibration, use a pH standard close to the pH of the solutions you wish to prepare. For a two point (slope) calibration, use two standards that flank the desired pH value.

6.4 A common source of error

Over 15 years ago, Illingworth (1981) noted a problem that was specifically attributable to poor behaviour (usually due to incorrect care) of the ceramic plug used to connect the test solution to the reference electrode. This plug itself can generate a significant potential which leads to errors in pH measurement. Moreover, the two-point calibration (Section 6.3) does not show or compensate for this error, because the standard buffers used to calibrate the electrode and meter are all at similar ionic strengths, and this error only makes itself felt when measuring the pH of solutions at very different ionic strengths. With so many people making '10 X' stock solutions of buffers, they will often be adjusting the pH of a solution of very high ionic strength. Under these circumstances, the error becomes important, but you will have no way to know about it.

Different electrodes vary considerably in their susceptibility to this effect, and what is needed is a simple test to determine the behaviour of your electrode. Illingworth suggests a simple test for ceramic plug artefacts:

◇ Preparation of concentrated stock buffers is not very good practice, as it is difficult to compensate for all the dilution effects. For pH–sensitive applications, this practice should be avoided.

- Measure the pH of the equimolar standard buffer (National Bureau of Standards buffer 5; Appendix 2, Buffer B). It should read pH 6.865 at 25°C.

- Now dilute the same standard buffer with 9 volumes of pure water (a 10-fold dilution). The meter should now read pH 7.065 ± 0.01 at 25 °C. Any discrepancy is due to liquid junction effects at the ceramic plug.

◇ Of course, we expect the pH to change because the buffer has been diluted, and the ionic strength is therefore lower. This is taken into account by the shift in pH we expect. Any discrepancy from this shift is due to electrode aberrations.

If the discrepancy is significant, it may be possible to restore or replace the creamic plug, but it is more likely that you will have to replace the electrode.

By now you will appreciate that the measurement of pH is not as straightforward as might be thought. The apparent precision of modern pH meters, with displays reading to three places of decimals, can give a completely unjustified and false sense of confidence. All of the potential for precision of measurement is there but ultimately, accuracy depends on attention to detail and a knowledge of potential pitfalls.

7. Alternative electrode technologies

A quick scan through equipment catalogues will prove to you that the glass electrode, despite its vintage, is the preferred sensor for determination of pH. It comes in a multiplicity of forms to meet almost all requirements of pH measurement and there are hundreds of varieties. But, the glass electrode is not without problems:

- The glass membrane is thin and fragile.

- The electrode has a restricted pH range and can interact with ions other than the proton.

- It is large and its bulbous nature makes it difficult to measure pH in small volumes or on surfaces.

- It is easily contaminated by biological fluids and solutions, particularly those containing protein and lipid. It is impractical for use *in vivo* or implantation.

- The electrical characteristics are not ideal and the measuring equipment to which it must be attached is necessarily sophisticated. Although this is a less serious impediment with the introduction of cheap microelectronics, the electrical connections to the electrode are still problematic.

7.1 How can we improve on the glass membrane?

One improvement comes from development of special types of glass. Most commercially available electrodes now cover the complete pH range for normal laboratory use (0–14) and will operate to high temperatures (100°C). Tough glasses are also available for industrial applications, but to improve in strength usually requires a compromise on pH range. Perhaps the most noticeable improvement in the ruggedness and general mechanical stability of the glass electrode has been the introduction of combination electrodes protected by plastic or epoxy resin sheaths. These sheaths cover the electrode stem and usually have an extension beyond the glass bulb. They are the preferred choice except in situations where certain noxious organic solvents are present (check with the manufacturer) or where very small volumes are to be measured. So tough are these devices that some suppliers recommend their use as stirrers as well as electrodes!

High concentrations of some common ions can interact with the glass electrode and give false readings. Among the ions of importance in biology are sodium and the ubiquitous Tris buffer ion. If you must use Tris, use Tris-insensitive glass electrodes. These, needless to say, are more expensive than the Tris-sensitive variety.

◇ These electrodes are also called 'Tris-competent' electrodes.

7.2 How can difficult or small samples be measured?

The development of the combination electrode has obviated one problem of electrode design—the need for bulky reference electrodes. Miniaturization and the provision of ultra-slim electrodes means that volumes of 1 ml or less can be measured in a suitably narrow vessel. The pH of surfaces may now be measured by the use of flat glass electrodes.

7.3 Resistance to contamination by biological fluids

This has been achieved largely by changes in mechanical construction. In the biochemical laboratory, where accurate measurement and control of pH are necessary, there is little that can be done apart from maintaining a strict regimen of electrode cleaning and replacement. However, in the field where absolute accuracy may be less essential, improvements have been made to electrode design. These include the plastic coating of the electrode body mentioned previously to improve strength, combined often with the provision of pointed electrode assemblies which can be inserted directly into soil and other materials. There are even electrodes incorporated into a knife blade which can be inserted into meat for pH measurement.

Another problem in hostile environments is the leakage of electrolyte from the assembly and the leakage of contaminants into it. This can be partially overcome by stabilization of the electrolyte as a gel within the electrode body. Protein, as well as sticking to the glass bulb and slowing electrode response, can also be precipitated by reaction with silver ions at the porous plug linking the test solution to the reference electrode. A stabilized gel electrolyte also helps to avoid this process.

7.4 Electrical characteristics of the glass electrode

There is little that can be done to reduce the resistance of the glass electrode and increase its current-delivering capacity. Modest voltages at low current and high source resistance will always have to be measured. Solid state electronics has made this less difficult but a problem remains—the electrode has to be connected to the pH meter by a screened lead. This lead, and the electrode itself, can pick up induced voltages. The signal can also leak away from the lead through paths of lower resistance. Both of these effects will result in apparent pH drift. The electrode is vulnerable to electrostatic effects (display of erratic pH values) produced by cleaning the electrode with dry rather than moist tissue.

The problems associated with the lead can be obviated by shortening it. Another source of difficulty is the attachment point of the lead to the meter body and the routing of the signal within the meter. Circuit boards are always made of fibreglass and a unity gain amplifier based on the field effect transistor or similar technology is usually placed as close to the electrode socket as possible. Some pH 'probes' are now available which do away with the lead altogether by attachement of the electrode directly to the meter body, although these are sometimes cheap designs built for convenience rather than accuracy. Electrodes are also available in which a microelectronic amplifier is actually incorporated into the electrode and is therefore physically close to the signal. An alternative strategy, however, is to make the microelectronic device itself the pH sensor!

8. The solid-state electrode

The electrical problems with the glass electrode can be diminished by adding a solid-state input amplifier at the connection point to the pH meter or by putting the amplifier into the electrode body. But could the electrode assembly be replaced altogether with a pH-sensitive semiconducting device?

For this purpose the insulated gate field effect transistor (IGFET) has been used. First of all we need to consider briefly how the FET works. The FET consists of three electrodes—source, drain and gate. Moderate current flows between the source and drain in much the same way that it would between the emitter and collector of an ordinary transistor. The gate electrode is used to control the current flow between the source and drain. The difference is that the gate electrode of the FET is coated with a very thin layer of metal oxide (usually silicon oxide, SiO_2) of high resistance. Therefore very little current can be injected into the transistor through the gate. Instead, when a voltage is applied to the gate, the electrical field generated serves to control the current between source and drain. In this way, a very small current applied through the gate is amplified to a very large change in the current flowing between source and drain (Figure 4.4).

In an ion-selective FET (ISFET), the insulating layer covering the gate electrode binds a specific ion. For a solid state pH electrode this ion would be the proton, H^+. When the proton binds to the gate an electrostatic field is generated which modulates the current flowing through the transistor. This current is then a measure of the hydrogen ion concentration or pH.

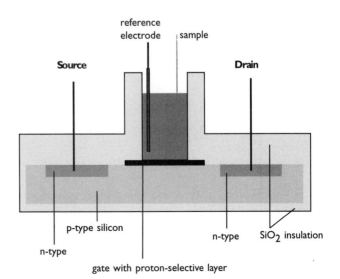

Figure 4.4

A simplified ISFET as a solid state pH electrode. The ion-selective layer controls the flow of current between source and drain. If the layer is sensitive to protons, then the device becomes a pH electrode.

The ion-selective layer used to date is silicon nitride (SiN_3) of less than 7 nm thickness, over an SiO_2 layer of 10 nm thickness. The output from these devices is typically about 56 mV/pH unit, which is close to the 59 mV/pH obtained theoretically for the glass electrode. These solid-state electrodes have a pH range of 1–13.

The gate electrode is small, typically about 0.5 mm square, and therefore very small drops of test solution can be applied to the gate making measurements on small volumes much easier. The electrode responds quickly (about 0.1 s), and is therefore useful for time-critical applications. It can also be miniaturized, is mechanically robust, and is useful in physiological and medical applications where it may be implantable. On the down side, SiN_3 has been shown to be thrombogenic, although other gate materials such as aluminium oxide (Al_2O_3) may be used.

◇ Thrombogenic: a tendency to initiate formation of a blood clot (a *thrombus*).

The electrodes are subject to a certain amount of drift. This is 0.6–3 mV/h for SiN_3 and about 0.3 mV/h for Al_2O_3. It will also be obvious that semiconductor devices are not normally expected to operate in an aqueous environment. As a result, the resistive layer will eventually separate on prolonged use in water. This problem may be circumvented to a certain extent by supplying the device with an insoluble vest of epoxy resin and silicone rubber, but this will ultimately deteriorate. The device therefore has a life of only a few weeks.

Very few solid-state devices of this sort are commercially available at the moment and they must be regarded still as specialist, research tools. However, one can envisage that the advent of the cheap, disposable electrode must be in view. This would bring the pH electrode into the realm of other disposable, solid-state biosensors.

Other ISFET devices have also been made, as have micro reference electrodes based on the FET. Micro devices incorporating an ISFET and Ag/AgCl reference system of as little as 0.3 mm diameter have been described.

Reading list

◇ A salutory lesson of the potential for one of the many errors in pH measurement.

Illingworth, J.A. (1981) A common source of error in pH measurements. *Biochem. J.* **195**, 259-262

5 Preparation of buffer solutions

- **Using written recipes**
- **Design of new buffers**
- **Practical considerations**
- **Describing buffers**

1. Where do I start?

Armed with a new understanding of buffer theory, you will, sooner or later, need to prepare a buffer for pH control. But how do you decide which buffer to prepare? Often, you will be able to follow specifications as presented to you in a laboratory manual or scientific paper. Beware—sometimes these recipes are incomplete or misleading. You will need your newly won capability to interpret these incomplete recipes, just as much as if you had to make the buffer from first principles. There will be times when you will need to design your buffer system from scratch. For example, you may need to design a new buffer for a chromatographic separation, determine the pH optimum of a system, or assess buffer-specific effects in a new procedure. In this chapter, we will discuss the practicalities of buffer preparation, and show how the rather abstract theory of the previous chapter really does influence 'wet' laboratory procedures.

2. Following existing recipes

In scientific papers or published protocols, you will read descriptions of buffers. In most instances, the description will (we hope) be complete and precise, and you will have no difficulties in making up such a buffer. Be forewarned, however, that there is usually more than one procedure that you can follow to make a buffer and you may have to choose the one that yields the same buffer as the original authors intended.

◇ This can require skills in mindreading in some cases!

Sometimes the published description will be vague. The scientific literature, and other people's notebooks are littered with 'shorthand' descriptions such as '0.1 M acetate, pH 5.0', '0.05M Tris/acetate', or, even worse, '2X Buffer A'—you will probably succeed in part if you attempt to prepare the first example, have difficulty with the second, and should not even attempt the last of these!

When you encounter a written buffer description, you should run through a quick checklist. First, establish exactly what the solution contains. As a description of a buffer, '0.1 M acetate, pH 5.0' leaves a lot to be desired. Obviously, you are told that the concentration of buffer ions is 0.1 M, but remember that in a buffer such as this, both the conjugate acid and the conjugate base are present. Thus, we are told that 'acetate' is 0.1 M, but does this really refer to the concentration of the acetate ions or to the sum of the concentrations of acetic acid and acetate? Usual conventions mean that the latter interpretation is correct, but be aware that we are making an assumption based on convention—someone who doesn't know the convention could inadvertently mislead you or be misled by you.

How else does this description fail? Nowhere are counterions discussed. It is impossible to prepare a solution that contains only acetate ions—considerations of electrical neutrality mean that every molecule of negatively charged acetate must co-exist with a positively charged ion. Usually, the buffer will have been been prepared with sodium or potassium ions, less commonly, ammonium ions, and bizarrely, Tris ions—whatever the ion used, it must be made explicit.

Lastly, the description 'pH 5.0' is incomplete. All buffers are temperature dependent, and a precise definition should include the temperature at which the buffer has that pH value. Whether designing your own buffers or following published recipes, an understanding of the theory will help you recognize the deficiencies in the description, and the need for precise and reproducible definitions. At the end of this chapter there are guidelines for complete specifications of buffers—try to adhere to them.

3. Designing a new buffer system

Major considerations in design of a buffer

❏ pH of buffer
❏ special properties (low UV absorbance?)
❏ buffer species
❏ buffer concentration
❏ ionic strength (extra salts?)
❏ temperature (working and preparation)
❏ solubility and stability of buffer species

Sometimes, you will need to design new buffers from first principles, and in the absence of a written definition or guidelines, you will have to make more decisions. You must consider and define variables such as pH, temperature, ionic strength, and the purpose to which the buffer will be put before you can design and prepare it.

3.1 At what pH is the buffer needed?

Because buffers give effective pH control within about one pH unit on either side of the pK_a value, the final pH of the buffer restricts your choice of buffer substances quite considerably. Use Appendix 1 to help you make this decision.

If you want to make a series of buffers that covers a range of pH values, you will, in all probability, need to prepare a set of buffers from different buffer substances, and with a considerable overlap in their pH ranges.

Lastly, you may be devising an experiment in which the pH is expected to change. Under these circumstances you must ensure that your buffer will behave as expected—maintaining the pH when required, but permitting the shift when it must take place.

3.2 Which buffer?

◇ Note that we are now discussing the value pK_a', the thermodynamically corrected and apparent pKa value.

Recall from Chapter 3 that a buffer resists pH changes best when the pH is equal to the pK_a' value, and the further that the pH is taken from this value, the less efficient the buffer is. Thus, the ability of a buffer to resist a pH change is governed by the displacement from the pK_a' value as well as the total concentration of buffer species. For the same buffering power, or buffer capacity, it would be necessary to use a higher concentration of buffer at a pH that is removed from the pK_a' than one would need when $pH = pK_a'$.

If your reaction is expected to undergo a significant uptake or release of hydrogen ions, then it would be preferable to use a buffer whose pK_a' is close to the pH you require. If you must use a buffer that is further removed from pK_a' higher concentrations will be needed.

Moreover, this effect is directional, in the sense that a buffer becomes weaker if the anticipated pH shift is away from the pK_a', but stronger if the pH shifts towards pK_a'. It is preferable to use a buffer at a pH greater than its pK_a' if the system generates hydrogen ions, as this will bring the pH closer to pK_a', and the buffer will become more effective. Similarly, if the experimental system consumes hydrogen ions, try to use a buffer that has a starting pH below pK_a'.

For any one pH value, there will usually be a choice of suitable buffer substances. How do you choose which buffer to use? Although buffer-specific effects can never be completely anticipated, there are a few points that you can consider. The goal is to design a buffer system that will have little or no influence on the system you are studying—it should not interact with the different components of your experimental system. Individual buffers have properties that render them less suitable for some applications, but more suitable for others. Appendix 1 highlights the most common problems that are experienced with individual buffers.

3.2.1 Charge

◇ The ionic strength of 0.1 M sodium phosphate at its pK_a is 0.2.

◇ The ionic strength of a 0.1 M sodium acetate buffer at its pK_a is 0.05.

The ionic strength of a salt solution is defined by a term that includes the square of the charge of the ionic species. Thus, polyvalent buffers such as phosphate, carrying multiple charges, will have a much higher ionic strength than monovalent buffers. Moreover, the ionic strength of polyvalent buffers will change more dramatically as the pH is changed.

3.2.2 Metabolic activity

Some buffers, for example, phosphate, glycine, or citrate, are also metabolically active, and may have unwanted effects in your system. Phosphate is a reactant in many enzyme and metabolic systems, and is a regulator (activator or inhibitor) in many others. Amino acid buffers, such as glycine, or metabolites such as tartrate or citrate are highly nutritious, and will support microbial growth in your solutions!

3.2.3 Insoluble salt formation

Some buffers, such as phosphate or carbonate, form insoluble salts with divalent metal ions such as calcium or magnesium, which will therefore precipitate out. If you need to include divalent metal ions in your system it will not be possible to use certain buffers.

3.2.4 Chelation of metal ions

Most common buffer species have very weak metal ion binding capacity, and it is common to see a chelator such as EDTA added to these buffers to prevent protein inactivation by heavy metal binding. Other buffers form complexes with metal ions that can reduce the effective concentration of these ions in solution. A buffer such as citrate or polydentate ligands such as amino acids can be good chelators of many metal ions, and may reduce their effective concentration in solution—important if the experimental system requires control of metal ion concentration. Moreover, because the conjugate acid and base will have different metal ion affinities, the ability to bind the metal ion will be pH-dependent. Detailed discussion of metal ion binding is beyond the scope of this book—if you are interested in metal ion effects you should already be aware of the complications that can be introduced by the buffer species.

3.2.5 Ultraviolet absorbance

◇ You will see, in Appendix 1, that the name 'Good' buffers refers to the scientist that developed them rather than being an implicit indication of quality—in fact, most of these buffers are really very useful for biological work.

Some buffers have a very strong near-ultraviolet (230–300nm) absorbance which can compromise their use in spectrophotometric procedures. Any buffer compound with an unsaturated or aromatic ring system will absorb quite strongly in the UV region. Low UV-absorbing buffers include many of the 'Good' buffers and the inorganic buffers (Appendix 1).

3.2.6 Buffers for chromatography

If you are preparing buffers for ion exchange chromatography, there is an additional complication. The column matrix, by definition, is charged; anion exchangers are positively charged and cation exchangers are negatively charged. The column matrix is therefore capable of binding ions of the opposite charge, and with some buffers this can cause problems.

For example, a phosphate buffer at neutral pH contains roughly equal amounts of the ions $H_2PO_4^-$ and HPO_4^{2-} An anion exchanger, such as DEAE cellulose or MonoQ beads, will carry a positive charge and, thus, the buffer ions are attracted to the matrix. But the HPO_4^{2-} will have a higher affinity for the matrix than $H_2PO_4^-$ and will be at a higher concentration near the column, which will produce a local increase in pH (Figure 5.1). The protein that binds to the column may therefore be surrounded by a solution that is at slightly different pH from the bulk solution pH of the buffer that is applied to the column. By the same argument, buffers with cationic species (such as the amine buffers) should not be used with cation exchange columns.

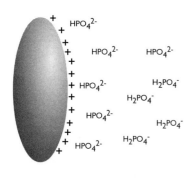

Figure 5.1

The charged surface of the anion exchange matrix attracts some buffer ions more than others, which leads to a local pH shift near the surface of the matrix.

How do we make sure that the buffer does not bind to the ion exchange matrix? Use a buffer species of the same charge as the chromatographic matrix. For a negatively charged matrix (a cation exchanger) we would need buffer species that are either uncharged or negatively charged, such as acetate.

3.3 What concentration of buffer?

The general answer to this question is 'as low as possible, as long as the pH can be controlled adequately'. But why should the pH change at all? Many reactions generate or consume hydrogen ions and, thus, in the absence of a buffer, would lead to a pH change in the reaction.

All solutions that are open to the air will absorb and dissolve carbon dioxide. Once dissolved, this forms carbonic acid and lowers the pH of the solution. A buffer must be at a sufficiently high concentration to resist this change.

It is hard to give exact guidelines but, in general, buffer concentrations of about 20–50 mM will provide effective buffering of pH over a range of $pK_a' \pm 1$. If your system is particularly prone to changes in hydrogen ion concentrations you might need much higher concentrations.

3.4 What ionic strength?

◇ Simple calculations will show that it will not be possible to obtain ionic strengths of 0.15 if you use most buffers at concentrations of 20 mM.

It is important that you make a decision about the ionic strength of the buffer, rather than simply ignoring this parameter, and allowing it to be whatever the buffer provides. Many enzyme systems prefer to operate in a solution that is isoionic (at the same ionic strength of most physiological systems), which is provided by an ionic strength of 0.15. As a first step, it might be wise to aim to adjust the ionic strength of all your buffers to this value, but explore the effect of ionic strength separately.

It is possible, in theory, to maintain the ionic strength by increasing the concentration of the buffer species to the appropriate concentration. This is undesirable for a number of reasons. First, you will be increasing the concentrations of complex solutes that might have undesirable effects. Secondly, if the pH of the buffer is ever changed, the ionic strength will also change substantially, because the only species controlling ionic strength are the conjugate acid and base. Finally, it can be expensive!

◇ Worked examples of buffer calculations are provided later in this chapter, but the software described in Chapter 6 will automate many of these steps.

It is preferable to maintain ionic strength by addition of a neutral salt, such as sodium chloride, to the buffer. You will need to know how much salt to add to bring the ionic strength to the required value, and therefore you must know how to calculate the contribution of the buffer components to the ionic strength as well. A low buffer concentration coupled with the use of potassium chloride or sodium chloride to sustain a relatively high ionic strength is recommended. The high concentration of ions of the monovalent salt will shield the system from the buffer ions, minimizing buffer-specific effects. Both of these monovalent salts are cheap, available in high purity forms, and have minimal specific effects.

There are some circumstances where a buffer must be as low an ionic strength as possible. One common example is the use of buffers to allow solutes to bind to an ion exchanger—the higher the concentration of ions in the solution, the weaker the electrostatic charge between the solute and ion exchange resin. In this case, care must be taken that the buffer can still sustain the pH, especially if the buffer is used to dissolve large amounts of proteins. Proteins can have quite strong buffering effects of their own, and shift the pH of weak buffers.

◇ A significant proportion of the buffering capacity of blood is provided by albumin and haemoglobin.

4. Examples of buffer design

There are three possible approaches to buffer design. First, you can stick only to established recipes, and follow them precisely. This is unlikely to be a satisfactory way to work. Sooner of later, you will need to design a new buffer from first principles, or need to analyse an existing buffer recipe in greater detail. A second approach is to prepare all the buffers that you need from published tables—the collection by Perrin and Dempsey (1974) is worth consulting, but many other texts are a useful source of such information. Lastly, you may prefer to design or re-design each buffer from first principles, taking responsibility for all aspects of the buffer design.

This section describes a series of scenarios in which you might find yourself. No specific buffer recipes will be given, but a few of the issues and pitfalls that you need to consider will be presented in each scenario.

4.1 Overall strategy

The flowchart in Figure 5.2 illustrates an overall strategy of buffer design. It cannot anticipate every nuance, because your specific needs cannot be predicted. However, note the branch points—a single buffer or a range of buffers, and the decision about control of ionic strength.

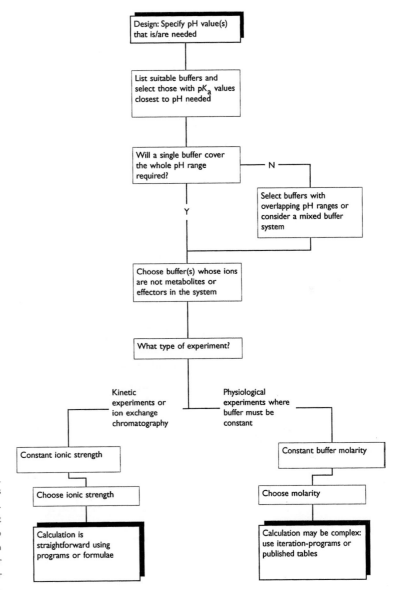

Figure 5.2

A flowchart that illustrates some of the steps in design of simple and more complex buffers. Specification of the buffers to be used is part of the experimental design. In addition to control of pH and ionic strength in an experiment, you may also need to consider 'downstream' stages, such as the need for separation in ion exchange chromatography.

4.2 A simple buffer with control of ionic strength by added salt

4.2.1 Introduction

This is a common buffer design, and will stand you in good stead for many applications. In this buffer system, the pH is maintained by a relatively low concentration of buffer species, and additional neutral salts are added to sustain the ionic strength at a value that you have specified.

4.2.2 Method

1. **Specify pH, temperature (T) and total ionic strength (I_{total}).**
2. **Choose a buffer with a pK_a near to the pH that you require.**
3. **Adjust pK_a for T and I.**
4. **Use the Henderson–Hasselbalch equation to calculate the concentrations of the conjugate base and acid.**
5. **Calculate I_{buffer} for this [acid] and [base] and remember to include counterions!**
6. **Calculate the concentration of neutral salt needed to make $I_{total} = I_{buffer} + I_{salt}$.**
7. **Prepare the buffer at T.**

4.2.3 Notes

This simple buffer assumes that you will prepare the buffer at the same temperature as you will use it.

If you add neutral salt, remember that I = [salt] if the salt is monovalent (NaCl, KCl), but for a polyvalent salt such as Na_2SO_4, $I = 3 \times$ [salt].

If you find that I_{buffer} exceeds I_{total}, you need to use a lower conentration of buffer.

4.3 Simple buffer with no external control of ionic strength

4.3.1 Introduction

It may seem at first glance that this is a simpler buffer than that described in Section 4.2, in which extra neutral salt was added to maintain a fixed ionic strength. However, this buffer is more complex, because there are two unknowns, pK_a' and I, that are interdependent. It is not possible to obtain both terms in one simple calculation, and an iterative calculation procedure is needed, to allow successive refinement of the values of I and pK_a'.

4.3.2 Method

1. **Specify pH and temperature (T).**
2. **Choose a buffer with a pK_a near to the pH that you require**
3. **Adjust pK_a for T.**
4. **Use the Henderson–Hasselbalch equation to calculate the concentrations of the conjugate base and acid.**
5. **Calculate I for this partition of conjugate acid and base, including counterions.**
6. **Calculate pK_a' from pK_a, using the calculated I value**
7. **Repeat the calculation of I, as in steps 4 and 5 above, but using pK_a'.**
8. **Refine the estimate of pK_a', using the newly calculated I value.**
9. **Repeat steps 7 and 8 until the correction to pK_a' is minimal (<0.001)—it is usually adequate to go through three cycles of refinement.**
10. **Prepare the buffer at T.**

4.3.3 Notes

This iterative refinement of pK_a' is tedious, and is better relegated to a computer program. Suitable software is described in Chapter 6.

It is assumed that the buffer is prepared at the temperature at which it will be used. Any deviation will mean that the buffer is incorrect.

4.4 A buffer prepared at a different temperature

4.4.1 Introduction

In many instances, buffers prepared at room temperature (for example, 20°C) are used at higher (37°C) or lower (cold room, 4°C) temperatures. It is necessary to calculate the effect of the temperature jump in design of such a buffer.

4.4.2 Method

1. **Specify pH_w at working temperature (T_w) and measure the preparation temperature (T_p)**
2. **Choose a buffer with a pK_a near to the pH that you require.**
3. **Use dpK_a/dT to adjust pK_a to T_w.**
4. **Use the Henderson–Hasselbalch equation to calculate the concentrations of the conjugate base and acid.**
5. **Make ionic strength decisions/corrections as needed (see Sections 4.2 and 4.3).**
6. **Use dpK_a/dT to adjust pK_a to T_p.**
7. **Keeping [acid] and [base] constant, calculate new pH_p at T_p from Henderson-Hasselbalch equation.**
8. **Prepare the buffer at T_p.**

4.4.3 Notes

This is appropriate for buffers prepared by the titration method (see Section 5). The calculation method specifies the pH at T_w, and as such, yields the concentrations of acid and base that are correct at T_w. However, if you prepare such a buffer at T_p, it will yield the wrong pH at T_w. This is one of those circumstances where you should trust the calculation.

The difference in pH between Tw and Tp can be calculated to be approximately $(Tw - Tp) \times dpK_a/dT$. Thus, you should be able to check that the corrections are correct.

This method makes no corrections for slight temperature effects on the activity coefficients of the buffers. In general these changes are so small that they can be ignored.

4.5 A series of buffers at different pH values using a single buffer species

4.5.1 Introduction

This type of buffer would cover a small pH range—within $pK_a \pm 1.0$. The method is simply a repeated set of calculations such as are found in Sections 4.2 and 4.3. The second method is based on mixing of the acidic and basic species in the ratios needed. Simple mixing of the two solutions will produce a series of buffers of different ionic strengths, and there will be slight errors due to the lack of compensation for salt effects on pK_a. If the ionic strength of the buffers is to be specified, then it will be necessary to adjust the buffer concentration, or to add neutral salts. The latter is preferred.

4.5.2 Method 1

1. **Select a buffer that spans the limited pH range.**
2. **Repeat the calculations in 4.2 or 4.3 as often as is needed.**
3. **Make T compensations as needed (section 4.4).**

4.5.3 Method 2

1. **Prepare solutions of the conjugate acid and base at the concentration that is needed.**
2. **Correct pK_a for T and I.**
3. **Use the Henderson–Hasselbalch equation to calculate the ratio of [base]/[acid].**
4. **Mix the solutions in the ratio specified.**

4.5.4 Notes

It is probably as simple to use the software described in Chapter 6 to calculate this set of buffers, which will be thermodynamically corrected.

4.6 A set of buffers covering a wider pH range, using different buffer species

4.6.1 Introduction

This set of buffers is often needed to span a wide pH range. There are several recipes for 'universal' pH buffers that span broad ranges, but we would not in general recommend them. First, although they sustain a good buffering capacity across the whole of the pH range that they define, there is considerable variation in ionic strength. Second, the buffer species used in these recipes may be incompatible with your system. We recommend that you decide on a range of buffers and ensure a) that they overlap considerably in the pH ranges they cover and b) that you sustain a constant ionic strength and a relatively uniform ionic environment by keeping the buffer species low, and adding monovalent salts to fix the ionic strength at a constant value.

4.6.2 Method

1. **Specify pH range, temperature (T) and total ionic strength (I_{total}).**
2. **Choose buffers that span the pH range that you require.**
3. **Adjust pK_a values for T and I.**
4. **Use the Henderson–Hasselbalch equation to calculate the concentrations of the conjugate base and acid.**
5. **Calculate I_{buffer} for this [acid] and [base] and remember to include counterions!**
6. **Calculate the concentration of neutral salt needed to make $I_{total} = I_{buffer} + I_{salt}$.**
7. **Prepare the buffer at T, or make the temperature adjustment as described in Section 4.4.**

4.6.3 Notes

Use each buffer for at least three pH values, ensuring at least a 0.5 pH unit overlap between different buffers. Discrepancy in the measured parameter will indicate buffer-specific effects.

The temperature and ionic strength corrections are best made by computer programs.

5. Practicalities of buffer preparation

There are two common methods of preparation of a buffer at a fixed pH. In both, the goal is the same—to prepare a mixture of the acidic and basic components in the correct proportions. In the first method, the two components are dissolved in the same solution, and the final pH should be the same as has been calculated (the 'calculation' method). In the other method, a solution of either the basic or the acidic component is dissolved, and the pH is adjusted, using strong solutions of acid or base and a carefully calibrated pH meter, to the required value (the 'titration' method).

In an ideal world, giving perfectly accurate pH meters, volumetric flasks, and balances, both methods will give exactly the same result. In practice, however, the two methods are less likely to be equivalent.

For highly accurate work, where the pH and ionic strength of the buffer solution are of paramount importance, it is preferable to mix the acidic and basic components in solution. It is not essential to check the pH of the solution under this circumstance, although you might wish to do so as a reassurance.

For many laboratory procedures, and provided that the pH meter is well cared for and properly calibrated (see Chapter 4), buffers made by titration can be equally acceptable, and accurate to better than 0.1 pH unit.

5.1 Which buffer compounds?

A quick look through the chemical catalogues will reveal that many buffer substances can be purchased in a bewildering variety of forms, and it is important that you are aware of the differences between these forms.

Many buffers are sold in both the acidic and basic forms. Tris hydrochloride and Tris base are different—one is the salt $TrisH^+Cl^-$, the other is the uncharged Tris base. If you dissolve the hydrochloride, it will release protons into the solution, and give a solution with a low pH. However, if you dissolve Tris base, it will pick up protons from solution, and the pH will rise above 7.0. This difference becomes significant when you adjust the pH of the buffer by the titration method, adding acid or alkali to adjust the pH.

With a solution of Tris base, you will need to add acid (such as hydrochloric acid) to lower the pH to the desired value. If you have prepared a solution of Tris hydrochloride, you will have to add a strong base, such as sodium hydroxide, to bring the pH to the desired value.

You may respond, 'But at the end of the procedure, I have still prepared a solution of say, 0.1 M Tris at the pH I need - so what's the problem?' The problem is this—*the two buffers are not the same*!

Let's look at these buffers more carefully, and consider the situation where we need to prepare a 0.1 M Tris buffer at pH = pK_a (in other words, the final concentrations of the acidic and basic species are equal).

If you start with Tris base, titrate to the correct pH with HCl, and then dilute to 0.1 M, you have prepared a solution that is 0.05 M Tris base (uncharged), 0.05 M Tris acid (charge = +1). But you cannot have a solution that is electrically unbalanced like this—what ions balance the positive charges on the Tris? The answer of course is chloride ions, added when the Tris base was titrated with HCl. So, we also have a solution that is also 0.05 M Cl⁻ ions. Taking all these species, the final ionic strength of the solution is:

$$I = \frac{1}{2}([\text{TrisH}^+] \cdot 1^2 + [\text{Cl}^-] \cdot 1^2)$$

$$I = \frac{1}{2}(0.05 \cdot 1^2 + 0.05 \cdot 1^2)$$

$$I = 0.05$$

Now, let's direct attention to the other Tris buffer, made by dissolving TrisH⁺Cl⁻, and then adjusting the pH with NaOH. In this instance, we start with 0.1 M TrisH⁺ and 0.1 M Cl⁻, but we must convert 0.05 M of the TrisH⁺ to Tris, by adding 0.05 M NaOH. In the final buffer therefore we have TrisH⁺ ions (0.05 M), Cl⁻ ions (0.1 M) and Na⁺ ions (0.05 M). Thus, the ionic strength is:

$$I = \frac{1}{2}([\text{TrisH}^+] \cdot 1^2 + [\text{Cl}^-] \cdot 1^2 + [\text{Na}^+] \cdot 1^2)$$

$$I = \frac{1}{2}(0.05 \cdot 1^2 + 0.1 \cdot 1^2 + 0.05 \cdot 1^2)$$

$$I = 0.1$$

So here we have two buffers, nominally the same, with the same concentration of buffer species, and the same pH, but one has an ionic strength that is twice that of the other! An understanding of the principles of buffer operation is essential to avoid these problems.

When making buffers by the titration method, always start with the buffer species that has the lowest charge. If using the calculation method, remember to include the salts in the final definition of ionic strength.

5.2 Water of crystallisation

Many salts can crystallize in different forms and many of the different forms vary in the number of water molecules that co-crystallize with them. Also, many buffer species can be purchased in hydrated and anhydrous forms. To illustrate, to obtain 0.1 mol of a salt that can crystallize with different numbers of water molecules, you would have to dissolve more of a hexahydrate than a dihydrate—in fact, an additional 7.2g. Why 7.2g?—because you will be adding an additional 4 mol of water (molecular weight=18) for every mol of buffer; 4 x 18 = 72g, and 0.1 mol = 7.2g. If you follow a recipe that was designed for the dihydrate, and then use the hexahydrate without compensating for the water of crystallisation, you will make the buffer incorrectly, and it will be less dilute than you planned.

5.3 Use of concentrated stock solutions

◇ Be aware that the correct preparation of a concentrated stock buffer is fraught with difficulties.

It is undoubtedly convenient to prepare buffers as stock solutions, and dilute them as required. A '10 X' buffer can provide for your needs for a considerable time, and has the additional advantage that you will be using the same stock solution for a large series of experiments, eliminating variability in the buffer composition.

A limitation in the preparation of stock buffer solutions can be the solubility of the buffer components. For any stock solution, each of the buffer species must be soluble. Of course, solubility must be maintained over the normal range of laboratory temperatures, and preferable, at cold room temperatures, for long term storage. It is not uncommon to see strong stock buffers precipitate during the winter, when night-time laboratory temperatures drop.

If the buffer has precipitated, it is still usable, provided that all of the precipitate has re-dissolved before a portion is taken. This is because the acidic and basic species may (and usually do) differ in their solubility, and thus, the precipitate will consist preferentially (or exclusively) of one buffer component. It follows that the solution now has a different ratio of acidic to basics species and, thus, will be at a different pH than if all of the precipitate was dissolved. You cannot simply take off the liquid phase that is sitting above the precipitate!

It is not appropriate here to give detailed lists of the stock solutions that can be prepared for biological buffers. Again, because the solubility of the acidic and basic components will differ, the maximum concentration of a stock solution that can be made is pH dependent. Extra solutes (such as sodium or potassium chloride) to control ionic strength, or which add a divalent cation (in the form of magnesium or calcium chloride, for example) can also have a dramatic effect on the solubility of the buffer species. Thus, if you want to prepare stock solutions, some experimentation, with relatively small volumes, is warranted.

◇ Of course, if you fail to make a stock buffer at '10 X' you can always add more water to make a '5 X', and so on—at worst, you will end up with a large volume of '1X'!

Stock buffers have other limitations. The most important of which is that they will undergo changes in pH when they are diluted, because the ionic strength will change. As we have discussed in Chapter 3, the working pK_a' of a buffer is affected by ionic strength according to the Debye–Hückel equation.

For buffers in which the charge on the acidic species is zero (e.g. acetate) or negative (e.g. phosphate) the working pK_a will increase as the buffer is diluted (the shift away from the true pK_a will diminish). Because of the simple relationship explained by the Henderson–Hasselbalch equation, the pH of the buffer will increase by the same amount:

For example, a stock acetate buffer diluted from 1 M to 0.1 M will increase in pH by about 0.4 units. The effect is even more dramatic with multiply charged buffers. These changes are significant, and the pH shifts (typically 0.5 pH unit) can have a dramatic effect on the proton concentration in solution, with consequent effects on biological processes. It is essential that stocks be designed correctly, and prepared properly.

The trick is to prepare the stock solution so that it is at the correct pH when it is diluted. There are a number of ways that you can do this, which depend upon the method that you used to make the stock in the first place. If you used the calculation method and made the stock solution by mixing the acidic and basic component in the first instance, you should calculate the recipe out at the working concentration, and then multiple the amounts of each component by the concentration factor. The stock solution will be wrong, but that is not important, as you will never use the buffer at this concentration.

If you prepare buffers by titration, then you must adjust the pH of the stock so that is correct when diluted. If an acetate buffer is diluted from 1 M to 0.1 M, the pH will increase by 0.43. Thus, the stock should be titrated to a value that is 0.43 units lower than that required after dilution. Since this buffer will normally be prepared from acetic acid and titrated with a strong alkali, the pH will be adjusted less than would be expected.

◇ In both methods, you should note that the pH correction will only be correct for a single dilution factor. A '10 X' buffer should only be diluted to '1 X',— if it is diluted to '2 X' the pH and ionic strength will both be incorrect.

Stock solutions contain much higher concentrations of the solutes, and as such, can have some specific effects. Strong solutions have a well-known tendency to 'creep' out of bottles, by a process of repeated dissolution and crystallization—the ability of potassium chloride to 'crawl' out of electrode wash vessels is a common illustration of this property. It is essential that the stock solutions be stored in well stoppered bottles.

Strong salt solutions (which is what most stock buffer solutions are) are also able to dissolve metals and silicates from containers, spatulas etc. It is probably better to store these solutions in plastic bottles.

◇ A common source of innoculum for such solutions is the ubiquitous wash bottle—which sits on a sunny bench every day, is repeatedly replenished, and can sometimes be seen to be sporting an attractive green tinge of algae on the bottom. This is an excellent source of biochemicals, but you may not want these in your experiments!

Strong buffer stocks can often be so hypertonic that they will not sustain growth of fungi, bacteria, or algae. A more dilute solution can generate conditions that favour growth. If you see any evidence of 'nasties' in a buffer solution, you really should discard it.

5.4 Handling buffer substances

If you wish to prepare a buffer by mixing two solutes, or by titration, ensure that you have the buffer components and strong acids/bases that you need. You will also need good quality water, prepared by a two stage purification comprising of two of reverse osomosis, distillation or polishing though cartridge purificaton systems. The water should be filtered through a 0.2 μm filter to diminish bacterial contamination.

Start by dispensing the buffer components. Mostly, this will mean weighing out the required amount, but some buffer species, such as acetic acid or ethanolamine, are liquids and will need to be dispensed differently. To obtain the required weight of these materials you will need to know the density of the liquid (densities of common buffer components are given in Appendix 1), so that you can dispense the required volume. Also, you might be able to weigh out the required weight of liquid on a balance, but be wary of liquids such as acetic acid or ethanolamine that give off unpleasant fumes.

Many solutes are hygroscopic, which means that they pick up water from the atmosphere. All solutes should be kept in tightly stoppered containers. If solutes are stored at 4°C or −20°C it is essential that you let the container come up to room temperature before you open the top—otherwise moist laboratory air will condense inside the container, and the contents will pick up water.

Weighing chemicals into weighing boats, or onto weighing paper introduces opportunities for 'fall-out' and loss of material as it is transferred into the final vessel. If possible, weigh the chemicals directly into the beaker or container that you will use to prepare the buffer. A balance that can be 'tared' is valuable, but make sure that you are not approaching the limit of the balance when you tare the vessel. If you have tared a 140 g beaker on a balance with a maximum capacity of 150 g, you can only weight 10 g of material into it. Obviously, plastic beakers are much lighter than glass.

◇ 'To tare' is to allow for the weight of the container. In this context, it means that the balance is reset to zero when the container is placed on the pan. This is relatively easy with electronic balances but more difficult with mechanical balances.

If at all possible, seek to avoid putting spatulas into bottles of buffer components. Many solutes can be tipped directly, with care, into a wide-mouthed beaker. If you overshoot, do not put the solute back into the stock bottle—assume it has been contaminated and throw it away. If you must put a spatula into a bottle, try to use one made of an inert plastic such as PTFE. Wash it in pure water, and wipe it dry on clean tissues immediately before use.

Many solutes are fine powders that float around, or aggregate into large lumps that come crashing out of the bottle. In either instance, they can create fall-out around the balance or on the pan. You must clean this up immediately. Apart from regulatory requirements and good laboratory practice, you are the only person who knows the nature of the anonymous powder and, therefore, the correct procedures for clean-up and safe disposal.

◇ 'Later' is not good enough—do it immediately, and do not let anyone else use the balance until it is clean.

You may be adding other solutes, such as EDTA or $CaCl_2$. Add these before you complete the buffer, not after it has been titrated or adjusted to the final volume. All solutes have a finite volume of their own, and will modify your buffer properties. Similarly, solutions of salts, such as EDTA, will contribute to the ionic strength of the solution, and thus, modify the pK_a— the pH will consequently change.

◇ You will have ensured that the pH meter is properly calibrated immediately beforehand.

Dissolve all of the solutes in a volume of water that is about three quarters of the final volume that you need. Then, if you are using the titration method, adjust the pH with strong monoprotic base or alkali (such as HCl or NaOH). Because you will be adding liquid solutions of acid or base, you must leave 'space' in the solution—hence the suggestion of about three quarters of the final volume. If all buffer components are added, as in the calculation method, make up the volume to the final value.

Sometimes, you may find that the buffer components do not dissolve immediately, although you are following a published recipe and would therefore expect them to be soluble. It is possible that the form of the buffer you are trying to dissolve is initially present at a concentration that exceeds its solubility. When you start to adjust the pH you may find that the insoluble material dissolves. Of course, as the material dissolves, the pH will drift, and you must make the final pH adjustment only when all components have been dissolved.

◇ Never try to adjust a buffer in a tall measuring cylinder with a stirrer bar at the bottom and electrode at the top.

Titrate the buffer to the correct pH using stock solutions of acid or alkali—typically 10 M and 1 M solutions of HCl and NaOH. The 1 M solution is more useful for adjustment of the pH of a dilute buffer, and the 10 M solution for concentrated or stock buffers. Add the acid or alkali slowly, dropwise into a well-stirred solution. Do not stir the solution with the pH electrode! After each addition, allow time for the titrant to equilibrate through the solution and, for strong buffers, for the electrode to stabilize in its response, before making further pH adjustments. Rapid mixing is essential, and the buffer will be mixed more efficiently if the vessel is wide and squat, rather than tall and thin.

Carelessness, or the use of titrant that is too strong relative to the buffer, will inevitably lead to overshooting. If you overshoot in the pH adjustment, you will have to start again. Repeated additions of acid and alkali (HCl and NaOH, for example) have the net effect of adding more salt to the buffer (in this case, NaCl). The pH should only be adjusted once, from the starting value to the desired value.

6. Correct buffer descriptions

When you make notes on your buffers, you should describe them in such a way that allows other scientists to prepare the exact equivalent of your buffer, even if they use a different method. A full buffer description will include the total concentration of buffer species, the temperature at which the buffer will be used, and the pH at that temperature. If salts were added, these should be specified also. Thus, a description such as '0.02 M acetic acid/sodium acetate buffer, 0.14 M NaCl, pH 4.20 @ 30°C ' is concise and unambiguous. Even if the reader prepares the buffer at a different temperature, they will know what the pH is at the working temperature and can make an appropriate correction. Similarly, a description such as this implies that the compensations were made for ionic strength effects. A precise reading of the method might even imply that the salt was added before the pH of the buffer was adjusted, if prepared by the titration method.

Further reading

Perrin, D.D. and Dempsey, B. (1974) *Buffers for pH and metal ion control.* Chapman & Hall, London.

6 Automatic buffer calculations

- **Software for buffer design**
- **Apple Macintosh**
- **World Wide Web**
- **MS-DOS/Windows**

1. Automating buffer calculations

In previous chapters, we have covered the theory and practice of buffer design. At this stage, you should be able to make value judgments about the most appropriate buffer for your purposes, and produce a complete recipe that gives you control of pH and ionic strength, and which is correctly adjusted for the temperature at which you will use the buffer. The series of case histories already covered will give you some idea of the manipulations that are needed.

You will have noted that most of these calculations are the same, irrespective of the buffer that is being used. Thus, such calculations are ideally suited to incorporation in a computer program that automates the task of design of a buffer. We have written such software that is used in our laboratories, and we make this software freely available to readers of the book, although it is not possible to include the software with the book itself. Detailed instructions for acquisition of this software will be found at the end of this chapter.

The software that we offer has not been written specifically to accompany the book, but provides the rigour embodied in the calculations that have already been discussed. Three platforms are covered—a stand alone MS-DOS program running in the Windows® environment, a 'non-windows' DOS program and, an Apple Macintosh® Hypercard® program that has evolved from software already described in the literature. Finally, we include description of a program that is accessible over the World Wide Web, precluding the need for any software other than a Web browser. Short descriptions of the software follows.

2. Macintosh Hypercard 'BufferStack'

This software was written under an early version of the Hypercard programming environment, and uses a 'stack' of cards as a metaphor. It has been upgraded and improved to run under Hypercard® version 2 and operates quite well using the Hypercard Player software supplied by Apple Computer®. Each buffer (Hepes, phosphate, acetate, and so on) occupies a single card, and all of the calculations for that buffer are

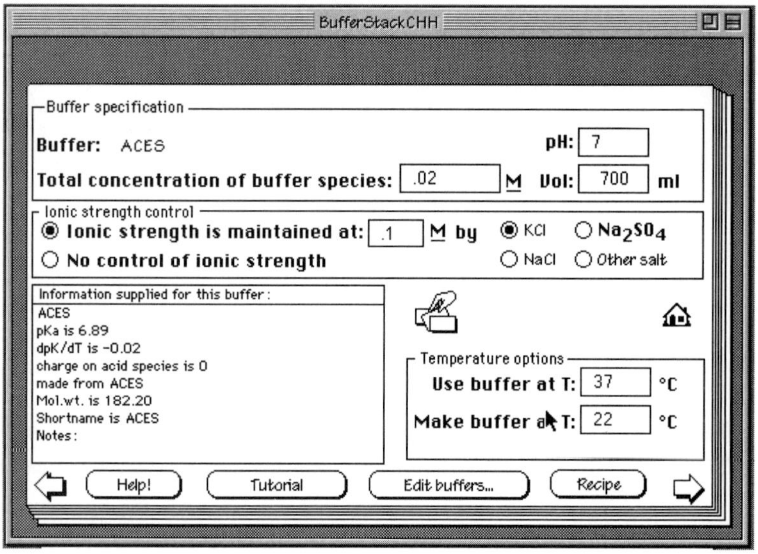

64 ◆ Automatic buffer calculations

Figure 6.1

A screen snapshot from the Hypercard program that calculates buffer recipes. The program comprises a set of cards, one for each buffer. The fields are completed, the options selected using the 'radio buttons', and the 'Recipe' button is pressed to calculate the information that is needed.

performed on that card. Figure 6.1 is an example card from the stack. The user completes the fields for pH and concentration, invokes options for control of ionic strength and elects, if needed, to prepare the buffer at a different temperature to that at which it will be used. The software then produces a complete recipe that specifies the weight of buffer component to be used. Figure 6.2 is an example recipe of the type produced by the software. BufferStack assumes that the buffer is prepared by titration. If you use BufferStack, therefore, you should ensure that your pH meter is accurate and properly calibrated—refer to Chapter 4 for details of electrode and pH meter care. The buffers made using this program will only be as good as the pH meter that you use. It is easy to add extra buffers, using the 'Edit buffers...' button at the bottom of the screen. Buffer information should be entered exactly as written in the information field—the program automatically extracts the correct data from this description.

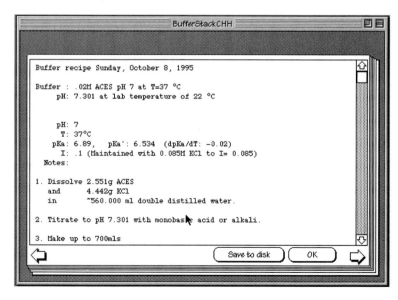

Figure 6.2

A screen snapshot from the Hypercard program after a recipe has been calculated. Note that the software allows a buffer to be prepared at a temperature different from that at which it will be used.

3. Buffer calculations on the World Wide Web

The buffer design that is embodied in the Hypercard Macintosh stack is also available over the Internet through the World Wide Web. To use this option, point your Web browser (such as Netscape®) to the following site:

◇ If you use the buffer software a lot, you make want to make a bookmark to this site.

```
http://www.bi.umist.ac.uk/buffers.html
```

After an introductory page, your browser will display a form similar to that shown (Figure 6.3) in which you can enter the data that define your buffer. When the data are entered, and the 'Submit' button is pressed, the data will be passed to a program at the site that will calculate the correct buffer recipe and return a formatted Web page that includes the recipe. The recipe can then be printed or saved as a text file.

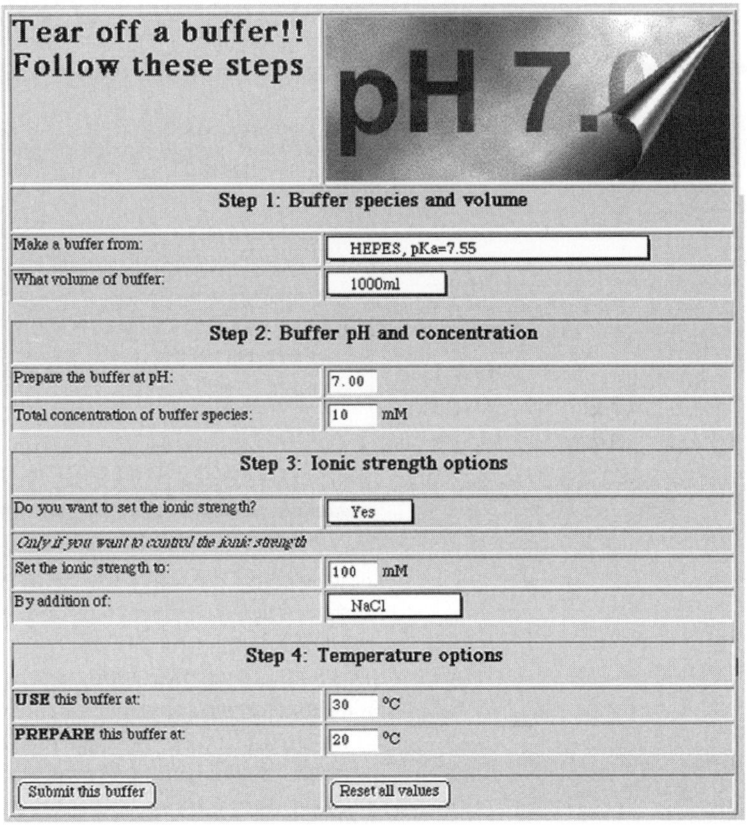

Tear off a buffer!! Follow these steps — pH 7.0

Step 1: Buffer species and volume

Make a buffer from:	HEPES, pKa=7.55
What volume of buffer:	1000ml

Step 2: Buffer pH and concentration

Prepare the buffer at pH:	7.00
Total concentration of buffer species:	10 mM

Step 3: Ionic strength options

Do you want to set the ionic strength?	Yes

Only if you want to control the ionic strength

Set the ionic strength to:	100 mM
By addition of:	NaCl

Step 4: Temperature options

USE this buffer at:	30 °C
PREPARE this buffer at:	20 °C

[Submit this buffer]   [Reset all values]

Figure 6.3

The design page of the buffer Website. Enter the information as requested in the form, and the buffer recipe will be returned after the 'Submit' button has been pressed.

4. MS-DOS/Windows software

◇ You may also specify any other buffer for which you know the pK_a, charge and temperature coefficient.

A further set of software has been written for MS-DOS/Windows environment. This software comprises a stand-alone application and use predominantly the 'calculation' method of buffer preparation, whereas the previous two programs focus on the titration method. The buffer is chosen from a pull-down menu (Figure 6.4) and the buffer conditions from a second menu item.

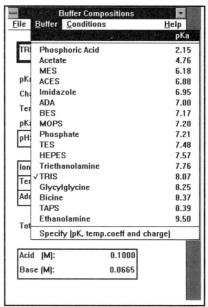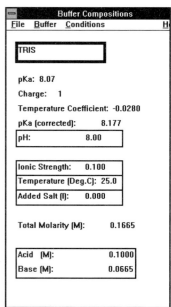

Figure 6.4

The Windows software for buffer calculations. This screenshots shows the range of buffers and the outcome of a request for a buffer

5. Obtaining the software

◇ Authors' e-mail addresses:
r.beynon@umist.ac.uk
jse@liv.ac.uk

◇ Authors' web sites:
http://www.bi.umist.ac.uk/buffer.html
http://www.liv.ac.uk/~jse/software.html

The software is freely available from the authors. If you have difficulties downloading the software, or to discover the current iteration of the software and the current site from where it may be downloaded, send us electronic mail .

There are no restrictions on the use of this software, other than the requirement that copyright remains with the authors and that the software is not re-distributed for gain. While we have tried to make sure that the software is as accurate as possible, we make no guarantees, express or implied, nor will we assume responsibility for the outcome of the use of any buffer designed with this software.

Futher reading

Beynon, R.J. (1988) A Macintosh Hypercard stack for calculation of thermodynamically-corrected buffer recipes. *Comput. Appl. Biosci.* **4**, 487-490.

AI Properties of common buffers

1. How to use this Appendix

This Appendix is a collation of many of the buffer compounds that you are likely to encounter in the biological sciences. In the detailed list, the buffers are arranged alphabetically by their trivial name, rather than by their pK_a values, because we will assume that you already know which buffer you wish to prepare. The complete set of buffers is listed alphabetically in Table A1.1. Table A1.2 lists the buffers according to their pK_a values, and you may consult this table first if you need to identify a series of buffers that you might use in that pH range. Figure A1.1 indicates the range of pH values that are covered by the buffers detailed here.

This appendix will not give instructions on the preparation of buffers, but indicates special applications, properties and hazards associated with the use of the compounds. Where no hazards are explicitly noted, it does not mean that you should assume the compounds are safe; it is your responsibility to assess hazards and act accordingly.

You may note that for some of the buffers (notably the 'Good' buffers) the pK_a values that we give are different to those commonly cited in catalogues and lists. This is because the other publications refer to a 'working' pK_a' calculated at an ionic strength of 0.1 and a temperature of 25°C. The values cited here are back-calculated to the thermodynamic pK_a values wherever possible. The set of 'Good' buffers are named after the scientist who first develped them (Good *et al.*, 1966) but can indeed, often be recommended as superior to more traditional buffer compounds.

For some, more complex buffers two chemical structures are often given. The top structure is the form of the buffer substance that is commonly purchased. The bottom structure is the ionic form of the buffer that is the conjugate acid at the pH at which the buffer will be commonly used. The functional group that ionises to give pH buffering is highlighted in the lower structure.

Further reading

Good, N.E., Winget, G.D., Winter, W., Connelly, T, Izawa, S., Singh, R.M.M. (1966) *Biochemistry*, 5, 467-477

68 ◆ Properties of common buffers

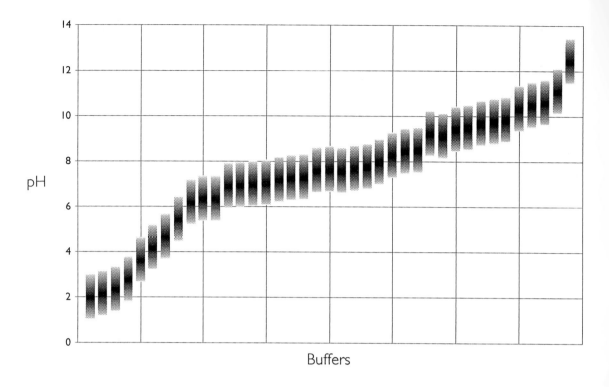

Figure A1.1

The pH span covered by the range of buffers described here. Notice that most of them cover the regon between pH 5 and pH 9 and that there is a large choice over the whole region. The shaded area approximates to the pH range covered by $pK_a \pm 1$ unit—the largest range one should normally consider for a buffer.

Buffer	pK_a	dpKa/dT
ACES	6.91	−0.02
Acetic acid	4.76	−0.0002
ADA	6.96	−0.011
Ammonia	9.25	−0.031
Benzoic acid	4.20	+0.018
BES	7.26	−0.016
Bicine	8.46	−0.018
Bis-tris	6.35	−0.02
Borate	9.23	−0.008
CAPS	10.51	−0.018
CAPSO	9.71	−0.018
Carbonate (pK_a1)	6.35	−0.0055
Carbonate (pK_a2)	10.33	−0.009
CHES	9.41	−0.018
Chloroacetic acid	2.88	+0.0023
DIPSO	7.71	−0.02
Ethanolamine	9.50	−0.029
Formic acid	3.75	0
Glycine (pK_a1)	2.35	−0.002
Glycine (pK_a2)	9.78	−0.025
HEPES	7.66	−0.014
HEPSO	7.91	−0.014
Maleic acid (pK_a1)	2.00	0
MES	6.21	−0.011
Methylamine	10.62	−0.032
MOPS	7.31	−0.011
MOPSO	7.01	−0.011
Phosphoric acid (pK_a1)	2.15	+0.0044
Phosphoric acid (pK_a2)	7.20	−0.0028
Phosphoric acid (pK_a1)	12.33	−0.026
Piperazine (pK_a2)	5.55	−0.015
Piperidine	11.12	−0.031
PIPES	7.14	−0.0085
Pyridine	5.23	−0.014
TAPS	8.51	−0.02
TAPSO	7.71	−0.02
TES	7.61	−0.02
Tricine	8.26	−0.021
Triethanolamine	7.76	−0.02
Tris	8.06	−0.028

Table A1.1

A summary of the buffers described in detail in the following pages. The buffers are listed alphabetically, as they are in the detailed descriptions that follow.

Buffer	pK_a	dpK_a/dT
Maleic acid (pK_a1)	2.00	0
Phosphoric acid	2.15	+0.0044
Glycine (pK_a1)	2.35	−0.002
Chloroacetic acid	2.88	+0.0023
Formic acid	3.75	0
Benzoic acid	4.20	+0.018
Acetic acid	4.76	−0.0002
Pyridine	5.23	−0.014
Piperazine (pK_a2)	5.55	−0.015
MES	6.21	−0.011
Bis-tris	6.35	−0.02
Carbonate (pK_a1)	6.35	−0.0055
ACES	6.91	−0.02
ADA	6.96	−0.011
MOPSO	7.01	−0.011
PIPES	7.14	−0.0085
Phosphoric acid (pK_a2)	7.20	−0.0028
BES	7.26	−0.016
MOPS	7.31	−0.011
TES	7.61	−0.02
HEPES	7.66	−0.014
DIPSO	7.71	−0.02
TAPSO	7.71	−0.02
Triethanolamine	7.76	−0.02
HEPSO	7.19	−0.014
Tris	8.06	−0.028
Tricine	8.26	−0.021
Bicine	8.46	−0.018
TAPS	8.51	−0.02
Borate	9.23	−0.008
Ammonia	9.25	−0.031
CHES	9.41	−0.018
Ethanolamine	9.50	−0.029
CAPSO	9.71	−0.018
Glycine (pK_a2)	9.78	−0.025
Carbonate (pK_a2)	10.33	−0.009
CAPS	10.51	−0.018
Methylamine	10.62	−0.032
Piperidine	11.12	−0.031
Phosphoric acid (pK_a3)	12.33	−0.026

Table A1.2

The same list of buffers as in Table A1.1, but sorted according to their pK_a values.

ACES

N-(2-Acetamido)-2-aminoethansulphonic acid
$pK_a = 6.99$; $dpK_a/dT = -0.02/°C$
$M = 182.2$ (free acid)

ACES is one of the 'Good' buffers and has a pK_a that is physiologically useful. The pK_a cited here has been back corrected to the thermodynamic value, and will therefore be different to the working pK_a' (6.90) value cited in the literature. ACES has a low UV absorbance at 260 and 280nm and is therefore a good buffer for chromatography. ACES is usually available as the free acid only.

Acetate

Ethanoic acid, acetic acid
$pK_a = 4.76$; $dpK_a/dT = -0.0002/°C$
$M = 60.05$ (acetic acid), $M = 82.03$ (sodium salt),
$M = 98.15$ (potassium salt)

Acetic acid is a commonly-used buffer at low pH values and is quite insensitive to temperature changes. Supplied as glacial acetic acid (d = 1.058 g/ml @ 20°C), which is very hygroscopic, and thus will absorb water and dilute itself unless tightly stoppered. A volume of 57.5ml of glacial acetic acid, made up to 1 litre with water, will yield a 1M solution. Because glacial acetic acid is so viscous, it is probably better to weigh it than to pipette it. Acetate salts are quite hygroscopic. Acetic acid is corrosive, and the fumes are an irritant.

ADA

N-(2-Acetamido)-iminodiacetic acid
N-(carbamoyl-methyl) iminodiacetic acid
$pK_a = 6.96$, $dpK_a/dT = -0.011/°C$
$M = 190.2$ (free acid)

Notes: ADA is another 'Good' buffer that has the properties of minimal physiological effects. The pK_a cited here has been back corrected to the thermodynamic value, and will therefore be different to the working pK_a' (6.62) value cited in the literature. Both carboxyl groups will be ionised at the pH in which ADA is commonly used, and thus, the conjugate acid has a charge of –1, and the conjugate base, –2. Thus, there is a stronger sensitivity to ionic strength effects with this buffer than with some of the other 'Good' buffers. ADA has some ability to bind metal ions.

Ammonia

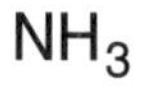

Ammonia

$pK_a = 9.25$; $dpK_a/dT = -0.031/°C$
M = 17 (ammonia), M = 77.08 (ammonium acetate),
M = 54.49 (ammonium chloride)

Notes: Ammonia has the advantage of being volatile, and as such, can be removed from solutes by freeze drying or on a centrifugal evaporator. It is often used as ammonium bicarbonate, which will buffer in the pH range 9-10, and which is fully volatile. An ammonia/ammonium hydrochloride buffer cannot be removed so easily by evaporation, and because ammonia is more volatile than HCl, will become acid during the drying process. Ammonia is supplied as a solution, or as ammonium hydrochloride—the latter, when titrated with NaOH, will yield a buffer containing NaCl, of course. Pure or strong solutions of ammonia are very unpleasant.

Benzoate

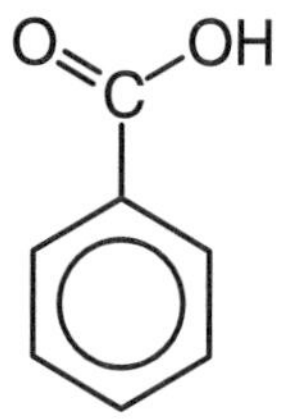

Benzoic acid

M = 122.1 (benzoic acid), M = 144.11 (sodium salt),
M = 160.22 (potassium salt)
$pK_a = 4.2$; $dpK_a/dT = 0.018/°C$

Notes: Benzoic acid has a strong ultraviolet absorbance which will make it unsuitable for many applications. It is relatively temperature sensitive.

BES

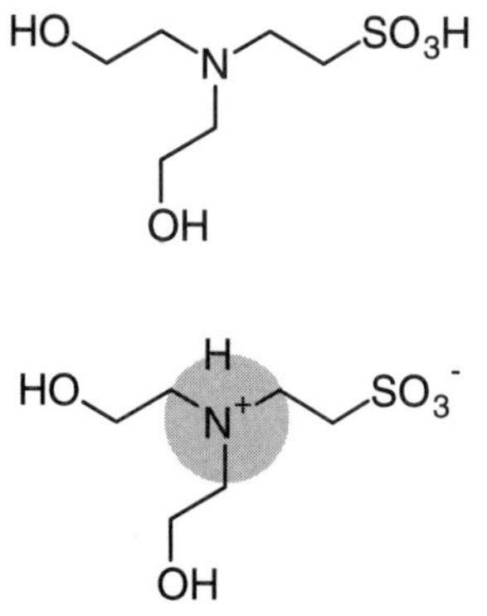

N,N-Bis-(2-hydroxyethyl)-2-aminoethanesulphonic acid

$pK_a = 7.26$; $dpK_a/dT = -0.016/°C$
M = 213.3 (free acid), M = 235.2 (sodium salt)

Notes: BES is a simple buffer that can also be obtained as the sodium salt and which therefore can be made by mixing solutions of the conjugate acid and base, or by titrating a solution of the acid with strong base. The pK_a cited here has been back-corrected to the thermodynamic value, and will therefore be different to the working pK_a' (7.15) value cited in the literature. BES has a very low ultraviolet absorbance at 260-280nm.

Bicine

N,N-Bis(2-hydroxyethyl)glycine
$pK_a = 8.46$; $dpK_a/dT = -0.018/°C$
$M = 163.2$

Notes: Bicine is a zwitterionic buffer.. In the region of pH 8, the carboxylate group is ionised, and thus, the charge on the conjugate acid is zero. The pK_a cited here has been back-corrected to the thermodynamic value, and will therefore be different to the working pK_a' (8.35) value cited in the literature.

Bis-tris

[Bis-(2-hydroxyethyl)imino]-tris-[(hydroxymethyl)methane]
2,2-Bis(hydroxymethyl)-2,2',2"-nitrilotriethanol
$pK_a = 6.35$; $dpK_a/dT = \sim -0.02/°C$
$M = 209.2$

Notes: Bis-tris is likely to be as temperature dependent as Tris buffer. The pK_a cited here has been back-corrected to the thermodynamic value, and will therefore be different to the working pK_a' (6.46) value cited in the literature.

Borate

H_3BO_3

Boric acid, H_3BO_3
$pK_a = 9.23$; $dpK_a/dT = -0.008/°C$
$M = 61.8$

Notes: Boric acid is not commonly used as a stand-alone buffer, but it finds application in Tris-borate buffers for electrophoresis in agarose gels. The ability of borate ions to bind to carbohydrates creates a 'borate front' that can assist in the electrophoretic separation. Boric acid is used as a primary pH standard (see Appendix 2).

CAPS

3-(Cyclohexylamino)-1-propanesulphonic acid
$pK_a = 10.51$; $dpK_a/dT = -0.018/°C$
$M = 221.3$

Notes: CAPS is a useful buffer at high pH values, and is used in some electroblotting applications. It has low absorbance in the 260-280nm ultraviolet range. The pK_a cited here has been back-corrected to the thermodynamic value, and will therefore be different to the working pK_a' (10.4) value cited in the literature.

CAPSO

3-(Cyclohexylamino)-2-hydroxypropanesulphonic acid
$pK_a = 9.71$; $dpK_a/dT = -0.018/°C$
$M = 237.3$ (free acid)

Notes: CAPSO is a relatively new buffer that is related to CAPS and CHES—all three buffers have good solubility and low UV absorbance at 260–280nm. The protonated form has a net charge of zero, and the deprotonated form has a charge of –1. The pK_a cited here has been back-corrected to the thermodynamic value, and will therefore be different to the working pK_a' (9.6) value cited in the literature.

Carbonate

Carbonate, bicarbonate
$pK_a1 = 6.35$; $dpK_a/dT = -0.0055/°C$
$pK_a2 = 10.33$; $dpK_a/dT = -0.009/°C$
$M = 105.99$ (Na_2CO_3), $M = 124.01$ ($Na_2CO_3 \cdot H_2O$),
$M = 84.01$ ($NaHCO_3$)

Notes: Because carbonic acid (H_2CO_3) in in equilibrium with H_2O and CO_2, (a reaction catalysed by the enzyme carbonic anhydrase), carbonate/bicarbonate buffers are usually employed to sustain CO_2 levels, as in tissue culture work. The carbonate ion is also used in volatile buffers such as ammonium bicarbonate. Because of the tendecy to exchange with CO_2 in the environment, carbonate buffers are not routinely used other than in specialist applications.

CHES

2-(Cyclohexylamino)-ethanesulphonic acid
$pK_a = 9.41$; $dpK_a/dT = -0.018/°C$
M = 207.3 (free acid)

Notes: CHES is from the same group as CAPS and CAPSO. All three buffers have good solubility and low UV absorbance at 260–280nm. The pK_a cited here has been back-corrected to the thermodynamic value, and will therefore be different to the working pK_a' (9.3) value cited in the literature.

Chloroacetate

Chloroacetic acid, Monochloroacetic acid, Chloroethanoic acid
$pK_a = 2.88$; $dpK_a/dT = +0.0023/°C$
M = 94.5 (free acid), M = 116.48 (sodium salt)

Notes: Chloroacetic acid is a stronger acid than acetic acid because of the electronegative chlorine atom drawing electrons away from the carboxyl group, thus reducing its ability to bind a hydrogen ion.

DIPSO

3-[Bis(2-hydroxyethyl) amino-2-hydroxy-propanesulphonic acid
$pK_a = 7.71$; $dpK_a/dT = \sim -0.02/°C$
M = 243.3 (free acid) M = 261.3 (monohydrate)

Ethanolamine

HOCH$_2$—NH$_2$

2-Aminoethanol, monoethanolamine
pK_a = 9.50; dpK_a/dT = –0.029/°C
M = 61.1

Notes: Ethanolamine is commonly used at moderately alkaline pH values. Being a primary amine, it will react with many amine modifying reagents, and is also rather temperature sensitive. Ethanolamine, as purchased, is a pretty smelly liquid (d = 1.02g/ml @ 20°C) and it may be preferable to make a stock solution at 2 M or 5 M in water. The pure liquid is harmful, and the vapour is extremely irritating.

Formate

H—C(=O)OH

Formic acid
pK_a = 3.75; dpK_a/dT = ~0
M = 46.03 (formic acid), M = 68.01 (sodium formate)
M = 84.12 (potassium formate)

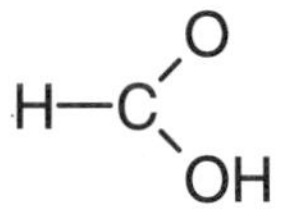

Notes: Formic acid is stronger than acetic acid, and has a very low temperature coefficient. Pure formic acid is a liquid (d=1.22 g/ml @ 20 °C). A volume of 38.5 ml of 98-100% formic acid, when diluted to 1 litre, will produce an approximately 1M solution. If you need to know the concentration accurately you will have to check it by titration. Buffers can be made by titration, or by mixture of the acid and the corresponding salt. Formic acid buffers are sometimes used in hplc. Formic acid is extremely corrosive.

Glycine

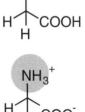

Aminoacetic acid
pK_a1 = 2.35; dpK_a/dT = –0.002/°C
pK_a2 = 9.78; dpK_a/dT = –0.025/°C
M = 75.1 (glycine), M = 111.53 (glycine hydrochloride)

Notes: Glycine is commonly used in electrophoresis systems. It is purchased as glycine (although the form shown in the top structure is chemically indistinguishable from the zwitterion, in which a hydrogen ion is transferred form the carboxyl to the amino group). Although some recipes make use of special properties of glycine (such as in electrophoresis systems) there is really little justification for using it as a general purpose buffer. As an amino acid, it has the potential to be a carbon and nitrogen source for microbial growth, it is a reactive primary amine, and it will find its way into samples that are destined for peptide sequencing, giving a glycine peak even when no glycine is expected!

HEPES

4-(2-Hydroxyethyl)piperazine-1-ethanesulphonic acid
$pK_a = 7.66$; $dpK_a/dT = -0.014/°C$
M = 238.3 (free acid), M = 260.3 (sodium salt),
M = 276.4 (potassium salt)

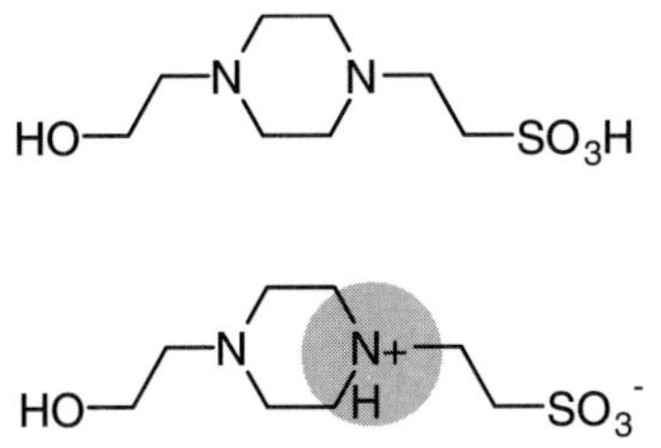

Notes: HEPES has a low absorbance in the ultraviolet, and is suitable for a wide range of applications. It has been used with great success in cell culture media. Buffers can be made by titration of the acid, or by mixture of the acid with the sodium or potassium salts.

HEPSO

4-(2-Hydroxyethyl)piperazine-1-2-hydroxypropanesulphonic acid
$pKa = 7.91$; $dpK_a/dT = -0.014/°C$
M = 268.3, but usually the monohydrate (M = 286.3)

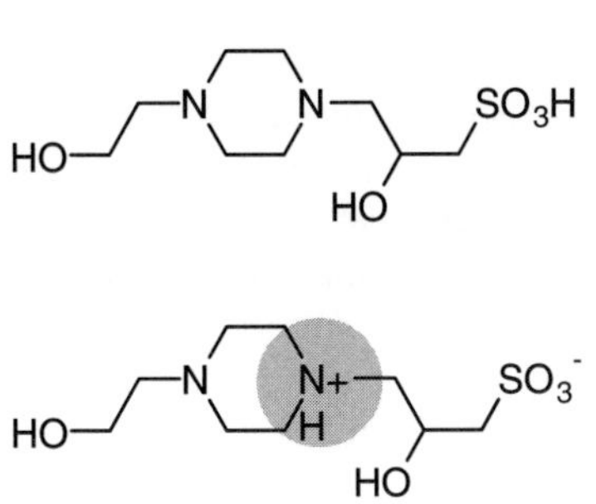

Notes: A very similar buffer to HEPES, with a slightly higher pK_a value. The pK_a cited here has been back-corrected to the thermodynamic value, and will therefore be different to the working pK_a' (7.8) value cited in the literature.

Maleate

Maleic acid
$pK_a1 = 2.0$; $dpK_a/dT = 0$
M = 116.1 (free acid)

Notes: Maleic acid is useful at low pH, although there are other buffers that are preferable. The –SH bonds of protein or free thiols can add across the double bond of maleic acid, and it is possible to produce proteins that have been irreversibly modified by this reagent—at the least they will have acquired a pair of negative charges, at the worst they will lose or change biological activity. Maleic acid is also an antimetabolite, inhibiting TCA cycle function. Not recommended in general—use formate, chloroacetate or phosphate in preference.

MES

2-(N-morpholino)ethanesulphonic acid
$pK_a = 6.21$; $dpK_a/dT = -0.011/°C$
M = 195.2 (free acid), M = 213.2 (monohydrate)

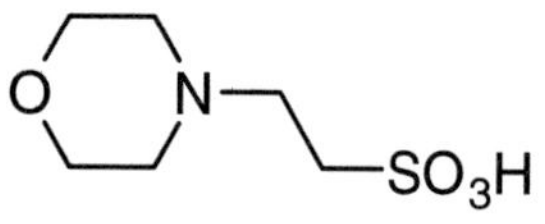

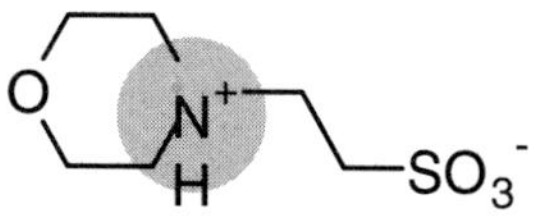

Notes: MES is another zwitterionic buffer, and the charge on the conjugat acid is net zero. Thus, although MES is ionic, the contribution to ionic strength is small.

Methylamine

Monomethylamine
$pK_a = 10.62$; $dpK_a/dT = -0.032/°C$
M = 31.1

Notes: Pure methylamine is pretty unpleasant, volatile, toxic and inflammable. If you must use this buffer, purchase it as a 40% solution (d=0.90g/ml, 20°C) which is a lot safer. Methylamine, as a primary amine, is very reactive. We recommend CAPS ($pK_a = 10.5$) as a safer alternative. Methylamine is only included here for the sake of completeness.

MOPS

3-(N-morpholino)propane sulphonic acid
$pK_a = 7.31$; $dpK_a/dT = -0.011/°C$
M = 209.3 (free acid), M = 231.2 (sodium salt),
M = 249.27 (sodium salt, monohydrate)

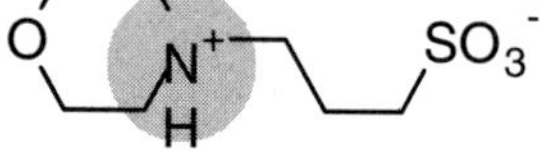

Notes: The pK_a cited here has been back-corrected to the thermodynamic value, and will therefore be different to the working pK_a' (7.20) value cited in the literature.

MOPSO

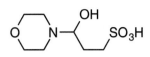

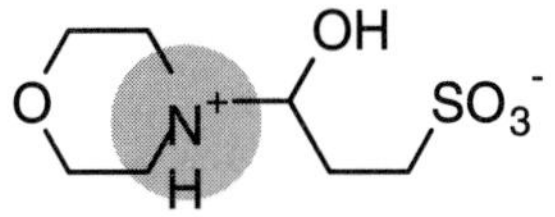

3-(N-morpholino)-2-hydroxy propanesulphonic acid, ß-hydroxy-4-morpholinepropanesulphonic acid
$pK_a = 7.0$; $dpK_a/dT = -0.011/°C$
M = 225.3 (free acid)

Notes: The pK_a cited here has been back-corrected to the thermodynamic value, and will therefore be different to the working pK_a' (6.9) value cited in the literature.

Phosphate

H_3PO_4
$\quad pK_a 1$
$H_2PO_4^-$
$\quad pK_a 2$
HPO_4^{2-}
$\quad pK_a 3$
PO_4^{3-}

Orthophosphoric acid
$pK_a 1 = 2.15$; $dpK_a/dT = +0.0044/°C$
$pK_a 2 = 7.21$; $dpK_a/dT = -0.0028/°C$
$pK_a 3 = 12.33$; $dpK_a/dT = -0.026/°C$
M = 98.0 (H_3PO_4), M = 142.0 (Na_2HPO_4), M = 268.07 ($Na_2HPO_4 \cdot 7H_2O$), M = 120.0 (NaH_2PO_4), M = 137.99 ($NaH_2PO_4 \cdot H_2O$), M = 174.2 (K_2HPO_4), M = 228.23, ($K_2HPO_4 \cdot 3H_2O$), M = 136.1 (KH_2PO_4)

Notes: Phosphate buffers are commonly used at neutral pH vales, but the other two ionisable groups can also be useful. Phosphate is a metabolite and an inhibitor of many reactions that involve phosphorylated metabolites. Phosphate buffers cannot be used in buffers that must also contain divalent metal ions such as calcium or magnesium, as the phosphates of these metals are rather insoluble.

Piperazine

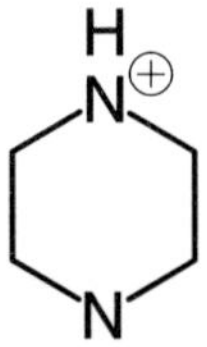

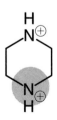

Piperazine
$pK_a 2 = 5.5$; $dpK_a/dT = -0.015/°C$
M = 86.14

Notes: Piperazine can cause burns. This is the base structure from which buffers such as PIPES were derived. These should be used in preference.

Piperidine

Piperidine
$pK_a = 11.12$; $dpK_a/dT = -0.031/°C$
M = 85.15, M = 121.61 (hydrochloride)

Notes: Piperidine is not really recommended as a buffer, as there are others at this pH range such as CAPS or phosphoric acid (pK_a3). Piperidine is an unpleasant liquid (d = 0.86 g/ml @ 20°C).

PIPES

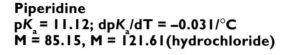

1,4 piperazine-bis-(ethane sulphonic acid), Piperazine-N,N'-bis(2-ethansulphonic acid)
$pK_a = 7.14$; $dpK_a/dT = -0.0085/°C$
M = 302.37 (free acid), M = 364.34 (disodium salt, monohydrate)

Notes: The pK_a cited here has been back-corrected to the thermodynamic value, and will therefore be different to the working pK_a' (6.8) value cited in the literature.

Pyridine

Pyridine
$pK_a = 5.23$; $dpK_a/dT = -0.014/°C$
M = 79.10 (free base), M = 115.56 (hydrochloride)

Notes: Pyridine is a liquid (d = 0.978 g/ml @ 20°C). It is totally unpleasant and toxic, and you need a good reason to use pyridine in preference to other buffers that cover the same range. Pyridine is used as a basic solvent for chemical reactions, but usually in the absence of water.

TAPS

N-tris[hydroxymethyl]methyl-3-amino-propanesulphonic acid, (2-[hydroxy-1,1-bis (hydroxymethyl)-ethyl]amino)-1-propanesulphonic acid

$pK_a = 8.51$; $dpK_a/dT = -0.02/°C$
M = 243.3; M = 265.3 (sodium salt)

Notes: The pK_a cited here has been back-corrected to the thermodynamic value, and will therefore be different to the working pK_a' (8.4) value cited in the literature.

TAPSO

3-{[tris(hydroxymethyl)methyl] amino}-2-hydroxypropane sulphonic acid
$pK_a = 7.71$; $dpK_a/dT = \sim -0.02/°C /°C$
M = 259.3 (free acid); M = 281.3 (sodium salt)

Notes: The pK_a cited here has been back-corrected to the thermodynamic value, and will therefore be different to the working pK_a' (7.6) value cited in the literature.

TES

3-{[tris(hydroxymethyl)methyl] amino}-ethanesulphonic acid
$pK_a = 7.61$; $dpK_a/dT = -0.02 /°C$
M = 229.25 (free acid)

Notes: Notes: The pK_a cited here has been back-corrected to the thermodynamic value, and will therefore be different to the working pK_a' (7.5) value cited in the literature.

Tricine

N-[tris(hydroxymethyl) methyl] glycine
$pK_a = 8.26$; $dpK_a/dT = -0.021/°C$
$M = 179.17$

Notes: The pK_a cited here has been back-corrected to the thermodynamic value, and will therefore be different to the working pK_a' (8.15) value cited in the literature. Tricine is one of the N-substituted glycines, and is related to bicine.

Triethanolamine

Triethanolamine
2,2',2"-nitrilotriethanol
$pK_a = 7.76$; $dpK_a/dT = -0.020/°C$
$M = 149.19$ (base), $M = 185.65$ (hydrochloride)

Notes: The base is liquid at room temperature (d = 1.124 g/ml @ 20°C). The fumes are unpleasant, and it may be preferable to make a strong stock, at say, 2.0 M and then keep cold and well stoppered.

Tris

Tris(hydroxymethyl)-aminomethane
$pK_a = 8.06$; $dpK_a/dT = -0.028/°C$
$M = 121.14$ (base), $M = 157.6$ (hydrochloride)

Notes: A ubiquitous buffer, that has several problems, including a high temperature sensitivity, reactivity as a primary amine, the need for Tris-competent pH electrodes and some undesirable effects on some biological systems. Its continued use may be more to do with familiarity and to published recipes than to scientific justification.

A2 Standards for pH calibration

1. How to use this Appendix

We list here three primary pH standards as recommended by the National Bureau of Standards. The solutions should be made up using freshly glass distilled water or glass distilled water which has had CO_2 removed by boiling and cooling. pH values are expressed to two decimal places which should be adequate for most practical purposes.

All of these standards have non-zero dpK_a/dT values, and therefore, the pH values delivered by the standards are temperature-dependent. As such, the standard value is quoted at a range of temperature values. It is essential that you calibrate the meter at the laboratory or solution temperature.

Nowadays, preparation of standards is simplified by pre-packaged buffer solutions, tablets or capsules, which, when dissolved in the correct volume of pure water, yield an accurate pH standard. These are often colour coded with dilute dyes.

Calibrate the meter using these standards in a one-point (buffer control only) or two-point (buffer and slope) calibration. Then, rinse the electrode thoroughly, to wash away the standard solution before inserting the pH probe into your solution. After you have measured the pH, rinse the electrode thoroughly again, and restore the electrode to the storage solution.

◇ It is most likely that you will use pre-prepared standard buffer tablets or solutions, but it is nonetheless useful to have recipes in case of emergencies.

1.1 Buffer A: Potassium hydrogen phthalate

This buffer is prepared at 0.05 M, by dissolving 10.212 g of potassium hydrogen pthalate in 1 litre of CO_2-free pure water.

1.2 Buffer B: Potassium sodium phosphate

This buffer comprises 0.025 M Na_2HPO_4 and 0.025M KH_2PO_4. Dissolve 3.533 g of Na_2HPO_4 and 3.388 g of KH_2PO_4 in 1 litre of CO_2-free pure water.

1.3 Buffer C: Sodium tetraborate

Sodium tetraborate (also known as borax) dissociates in solution to form equimolar concentrations of metaboric acid and the metaborate ion. This conjugate acid/base pair form such a reproducible [H⁺] in solution that it is acceptable as a primary standard. The standard is normally prepared at 0.01M, by dissolving 3.814g of the decahydrate, $Na_2B_4O_7 \cdot 10H_2O$ in 1 litre of CO_2-free pure water.

T (°C)	Buffer A	Buffer B	Buffer C
0	4.00	6.98	9.46
5	4.00	6.95	9.40
10	4.00	6.92	9.33
15	4.00	6.90	9.28
20	4.00	6.88	9.23
25	4.01	6.87	9.18
30	4.02	6.85	9.14
35	4.02	6.84	9.10
38	4.03	6.84	9.08
40	4.04	6.84	9.07
45	4.05	6.83	9.04
50	4.06	6.83	9.01
55	4.08	6.83	8.99
60	4.09	6.84	8.96
70	4.13	6.85	8.92
80	4.16	6.86	8.89
90	4.21	6.88	8.85
95	4.23	6.89	8.83
$\Delta pH_{0.5}(25°C)$	+0.052	+0.08	+0.01

Table A2.1

The pH values of three primary pH standards. Each of these buffers is intended to be used at the strength indicated. If the buffers are diluted, they change their pH. This is expressed as $\Delta pH_{0.5}$, the change in pH caused by a 1:1 dilution of each of the standards.

Glossary

Acid An entity that can donate a hydrogen ion (Lowry–Brönsted definition).

Acidosis A clinical condition in which blood pH is lower than normal.

Alkalosis A clinical condition in which the pH of the blood is raised above normal.

Base An entity that can receive a hydrogen ion (Lowry–Brönsted definition).

Beta value *See* buffer capacity.

Buffer A compound that resists a change in pH on addition of acid or alkali, by virtue of a compensatory adjustment of the acid–base equilibrium.

Buffer capacity A rarely-used term that reflects the ability of a buffer to resist addition of acid or base.

Glass electrode A proton sensitive electrode for pH measurement. Most designs incorporate a second electrode to provide a reference voltage.

Hydronium ion The hydrated form of the hydrogen ion (proton). Generally considered as H_3O^+, it is more likely to comprise additional water molecules.

K_a The dissociation constant for loss of a proton by an acid.

Mixed buffer A buffer made of a mixture of a weak acid and a weak base that are chemically different, such as Tris–acetate.

pD The equivalent to the pH scale, but in D_2O rather than H_2O.

pH standard A buffer that has a well defined behaviour and stability in solution and which can be used to calibrate a pH electrode and meter.

pK_a The logarithmic form of K_a, defined as $-\log_{10}K_a$.

Polyprotic buffer A buffer that has more than one ionizable group, and therefore, multiple pK_a values.

Weak acid An acid that has only partly dissociated into its deprotonated form under typical conditions of use.

Index

ACES 71
acetate, acetic acid 9, 15, 45, 71
acid dissociation constant 9
acidosis 2
activity, activity coefficient 25, 26
ADA 71
alkalosis 2
alternative electrode technologies 41
ammonia 72
apparent pK_a 32
automatic buffer calculations 63

β values 25
benzoate 72
BES 72
bicarbonate 74
Bicine 73
Bis-tris 73
blood 2
borate 73
buffer adjustment control, pH meter 39
buffers
   substances and selection 57, 67–82
   for chromatography 48
   concentration 49
   control of ionic strength by added salt 52, 53
   descriptions 62
   design 46, 50
   practicalities in design 57
   preparation at different temperature 54
   stock solutions 59
buffering capacity 25
butyric acid, butyrate 15

calibration, pH meter and electrode 39
CAPS 74
CAPSO 74
carbonate, carbonic acid 74
care of a pH electrode 36
charge 47, 59
chelation 48
CHES 75
chloroacetate, chloroacetic acid 9, 15, 74
chromatography 48
concentration, buffer species 49
conjugate acid 9
conjugate base 9
contamination of pH electrode 36

Debye–Hückel 30
design of a new buffer system 46
dichloroacetic acid 15
dilution effects 30
DIPSO 75

effect of temperature on pK_a 31
electrode
   calibration 39
   contamination and cleaning 36
   glass 35
   mechanical abuse 36
   replacement 37
   solid state 43
endosomes 2
equilibrium constant 19
error in pH determination 40
ethanolamine 76

formate, formic acid 9, 76

glass electrode 35
   calibration 39
   contamination and cleaning 36
   mechanical abuse 36
   replacement 37
glycine 76

handling buffer substances 60
Henderson–Hasselbalch equation 19
HEPES 77
HEPSO 77
hydrogen bonding 17
hydronium ion 6
hygroscopic 61
Hypercard® 63

inductive and electrostatic effects on pK_a 14
insoluble salts 48
ionic product of water 6
ionic strength
   definition 26
   formula 27
   influence on pK_a 29
   specification in buffer design 49
ionizable groups 14

Lowry–Brönsted 7
lysosomes 2

maleate, maleic acid 77
measurement of pH 35
mechanical abuse of pH electrode 36
MES 78
mesomeric effects on pK_a 15
metabolic activity of buffer compounds 47
methylamine 78
'mixed' buffers 33
MOPS 78

MOPSO 79
MS-DOS/Windows software 66

Netscape® 65
neutral pH, definition 5

PCR, polymerase chain reaction 31
pepsin 36
pH meter
   buffer adjustment control 39
   calibration 39
   description 35
   slope adjustment 40
pH
   pH scale 4
   physiological 2
   of a solution of a weak acid or base 11
   standards 39, 83, 84
phosphate 9, 79
phthalate, pH standard 83
physiological pH 2
piperazine 79
piperidine 80
PIPES 80
pK_a
   apparent 32
   effect of ionic strength 29
   effect of temperature 31
   role of structure in determination of 13–17
   thermodynamic 14, 20
polyprotic buffers 32
potassium hydrogen phthalate, pH standard 83
potassium sodium phosphate, pH standard 83
practicalities of buffer preparation 57
preparation of buffer solutions 45
propionate acid, propionic acid 5
proton, hydrogen ion 4
pyridine 80

recipe, buffer 45, 64
replacing a pH electrode 37

series of buffers, different pH values 55
slope adjustment, pH meter 40
sodium tetraborate, pH standard 84
software for buffer design 3, 64–66
solid-state electrode 43
solubility of buffer compounds 59, 61
standard pH solutions 39, 83, 84
statistical effects on pK_a 16
steric effects on pK_a 16
stock solutions of buffers 59

storage of pH electrodes   37
structure, effect on acidity   13

TAPS   81
TAPSO   81
TES   81
temperature, effect on pK_a 31

theory of buffer action   18
trichloracetic acid   9, 15
Tricine   82
triethanolamine   82
Tris   45, 82

ultraviolet absorbance by buffers   48

water of crystallisation   58
water, purity   60
weak acids and bases   7, 13
Windows®   3, 66
World Wide Web   3, 65